HUMAN BODY
FACTFINDER

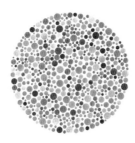

Created and Written by
John Farndon and Angela Koo for Barnsbury Books

Image Coordination
Ian Paulyn

Production Assistant
Jenni Cozens

Editorial Director
Paula Borton

Design Director
Clare Sleven

Publishing Director
Jim Miles

This edition published by Dempsey Parr, 1999
Dempsey Parr, Queen Street House, 4 Queen Street, Bath, BA1 1HE, UK

Copyright © Dempsey Parr 1999

2 4 6 8 10 9 7 5 3 1

Produced by Miles Kelly Publishing Ltd
Bardfield Centre, Great Bardfield, Essex CM7 4SL

ISBN 1-84084-517-1

Printed in Singapore

HUMAN BODY

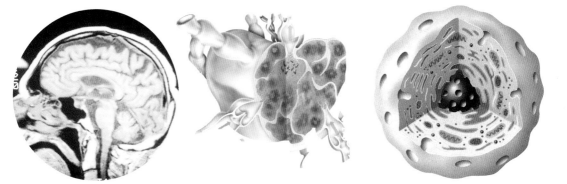

FACTFINDER

DP
DEMPSEY
PARR

INTRODUCTION

Humans are visual creatures.
We rely on eyesight more than
any other sense—especially to find
out about the world around us from
words and pictures. Lists of facts and
descriptions of events may contain
concentrated knowledge, but adding
illustrations helps to bring the subject
alive. They encourage us to delve
further, appreciate, and enjoy, as well
as retain the information.

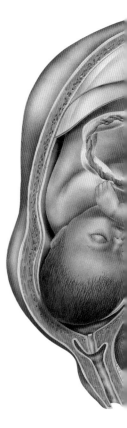

The FACTFINDER series is packed with a huge variety of facts and figures. It also explains processes and events in an easy-to-understand way, with diagrams, photographs, and captions. Fact panels on each main page area contain information for ready reference. Each title is divided into sections that deal with a major aspect of the subject. So look and learn, read, and remember— and return to again and again.

CONTENTS

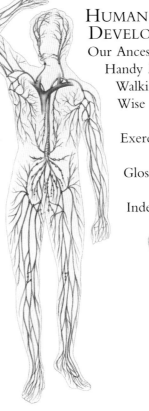

THE BODY MACHINE

Human bodies come in all shapes and sizes. Some of us are fat. Some are thin. Some are tall. Some are short. And we all look different—at least to other humans we do. But we are all made of the same materials and work in the same way.

Every moment of the day, even while you are asleep, your body is busy at work. Your chest is steadily pumping air in and out of your lungs. Your heart is squeezing away, sending blood shooting through your blood vessels. Chemical agents are at work in the intestine. Nerve signals are buzzing to and fro,

BODY POWER
Your body is made up of millions of cells, each quite different, but each forming part of a special body system. All the cells and systems in the body work together to sustain the remarkable living organism that is you.

speeding information from the senses to the brain and carrying instructions to all parts of your body. Chemical messengers called hormones are galvanizing all kinds of processes into action.

All of this whirl of activity is designed with just one purpose in mind—to keep you alive. Your body is made up of millions of tiny cells, each quite similar in essence to every living cell on earth, right down to bacteria. What makes you special and unique is the way your body's cells work together to keep you alive and healthy.

BODY POWER
A top athlete's body works at peak efficiency—but works in precisely the same way as yours.

Body facts
- The tallest man ever was American Robert Wadlow, (b. 1918) who reached 8 feet, 11 inches (272cm) tall.
- The shortest man ever was American "General Tom Thumb" who was just 3 feet, 4 inches (102cm) tall when he died in 1885.
- The fattest man ever was American Jon Minnoch, who weighed 1,397 pounds (635kg) when he died in 1983.
- The lightest human adult ever was Mexican Lucia Xarate, who weighed just 4 pounds, 11 ounces (2.13kg) when she was 17 in 1889.
- A woman in Africa is thought to be 150 years old, but no one can prove this.
- In 1998, Frenchwoman Jeanne Calment, born in 1875, died at age 123.
- The most children born at once were decaplets—2 boys and 8 girls born in 1946 in Bacacy in Brazil.
- Mr. and Mrs. Vassileyev, who lived near Moscow in 1782, had 69 of their own children.

STUDYING THE BODY

The first great strides in our knowledge of the human body came when great Renaissance anatomists such as Vesalius began cutting up corpses. Since then, each new technology—the microscope, X-rays, CT scans, and much more—has provided scientists with new and often startling insights into the way our bodies work.

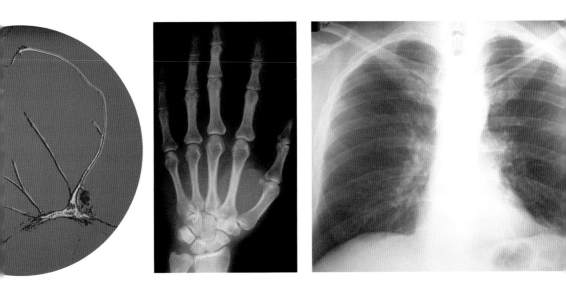

THE BODY IN HISTORY

Scientists look at the human body in two main ways. First of all they can look at its anatomy. Anatomy is the geography of the body. It looks at the structure of the body—just where everything is and how it all fits together. Secondly, they can look at its physiology. Physiology is just what goes on and how it all works. It looks at the processes

of its cells, tissues, organs, and systems. If anatomy is a still photograph, physiology is a movie.

Both branches of study date back at least to Ancient Egyptian times, and our understanding is the work of many centuries. Many anatomical terms date back to the Flemish scientist Andreas Vesalius, whose book of 1543 is the great classic of anatomy. Some even date back to the Roman physician Galen.

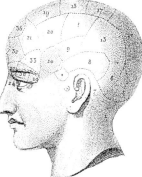

BUMPS ON THE HEAD
Some parts of human anatomy are not quite as people once thought. In the nineteenth century, many people studied bumps on the head to find clues to personality. This was called phrenology.

NAMING PARTS
Although there are often simple English equivalents, every part of the anatomy has an official name, which is usually in Latin. We owe this wealth of Latin names mainly to the anatomists of the fifteenth and sixteenth centuries, who were very familiar with Latin. Although the names are often hard to spell and remember, they are the same for people in every country.

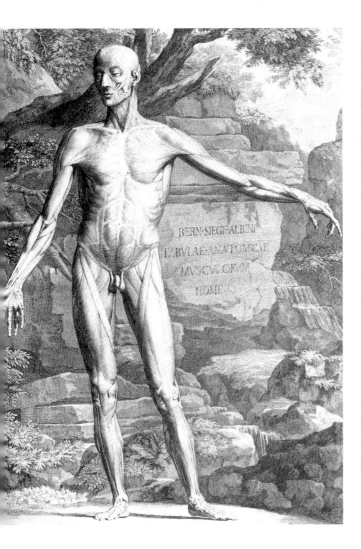

How the body is described

- When describing anatomy, most textbooks assume the body is standing upright, facing you with arms hanging down by the side.
- The body is then divided into two halves from front to back down the middle. The dividing line is the "median" plane.
- Anatomists then talk of the right and left of the body.
- The body is also divided in half crosswise. This dividing line is called the "coronal plane."
- Things in front of the coronal plane are said to be anterior or ventral.
- Things behind the coronal plane are said to be posterior or dorsal.
- You can use both Latin and English words to name parts of the anatomy, but it is simpler to use English, and this is what is done in this book. So we have called the pectoral muscles the pectorals, rather than pectoralis, for instance.

ANCIENT ANATOMISTS

The first doctor anyone knows about was Imhotep, who lived in Ancient Egypt 4,600 years ago. It is recorded that people traveled from far and wide to be treated by him, and after his death he was made a god. The first medical textbook comes from Ancient Egypt too, and in the Smith papyrus of 1500 B.C., it describes how to set broken bones, how to take the pulse, how to stop bleeding, and much more.

However, it is with the Ancient Greeks and Romans that our understanding of the body really began. The Greek thinker

ROMAN RAZORS
Until quite recently hairdressing and surgery were done by the same person, called a barber surgeon, who was good with knives.

COLISEUM
The coliseum was Ancient Rome's great sports arena. Here thousands of people would watch gladiators fight to the death—which provided plenty of material for doctors.

Empedocles (born about 492 B.C.) saw that the heart was linked to blood vessels, for instance. The greatest of the Roman doctors was Galen, who was born around A.D.130 and learned a great deal from cutting up animals. He wrote many books describing the skeleton, the muscles, and the nerves. People thought so much of his knowledge that for more than 1,000 years, doctors would consult Galen's books rather than look at a body.

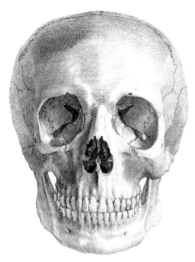

SKULL
Skulls have survived since
ancient times to give us some
idea of the very earliest
medical practices. It is from
old skulls, for instance, that
we know of an operation
called trephination, which was
performed at least 10,000
years ago. This involved
drilling holes right through the
skull and into the brain. No
one knows quite why it was
done, but it may have been a
magic ritual to release spirits
trapped inside the head.

THE FOUR HUMORS

Until a few hundred years ago many thinkers
believed that life had four qualities: hot,
cold, wet, and dry, linked to four elements
—fire, air, water, and earth. In the body,
these gave four "humors" or fluids: blood,
phlegm, yellow bile, and black bile. If any of
these qualities were out of balance in the
body, you would be ill.

Hippocrates

- Many doctors today practice
  what is called the
  Hippocratic code and may
  swear an oath, called the
  Hippocratic Oath.

- Under the Hippocratic
  Oath, doctors promise to
  be sensible, thoughtful,
  and modest.

- Doctors also follow the
  Principle of Confidentiality,
  which says that personal
  information about a patient
  should be kept secret.

- The Hippocratic Oath dates
  back to the Ancient Greek
  physician Hippocrates, who
  lived on the island of Cos
  from about 460 B.C.

- Hippocrates ran a hospital
  on Cos and taught a
  sensible, unsuperstitious
  approach to illness.

- Hippocrates insisted that
  doctors should only give
  treatment when it was
  definitely needed, and
  should always know the
  likely effects of any
  treatment before they
  applied it.

CUTTING INSIDE BODIES

For more than a thousand years, people tended to accept what the Ancient Roman physician Galen said about the body almost without question. Then in the fifteenth and sixteenth centuries, some thinkers realized it might be better to actually look at bodies themselves. So they began to obtain dead bodies from graveyards and started cutting them up carefully to see exactly how they were put together. This is called dissection.

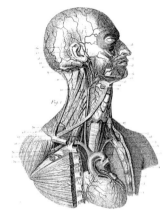

GRAVE SECRETS
It was only by cutting up corpses that scientists began to learn about how the body was put together in detail. But finding corpses for dissection was not always easy, and many anatomists resorted to snatching bodies from graveyards at the dead of night.

ILLUSTRATED FOOT
Anatomical drawings were often made by fine artists, including the great Italian painter Leonardo da Vinci (1452-1519), to ensure accuracy. This engraving of the bones of the feet is from the seventeenth century.

Piece by piece the dissections of physicians such as Andreas Vesalius built up a very detailed picture of human anatomy—the way the human body is put together. Gradually, physicians began to learn about physiology, too—the science of the workings of the body. They learned for instance, that the heart is a pump, and blood circulates around the body.

THE BODY REVEALED
Andreas Vesalius was the most famous of all the anatomists working at Padua in Italy in the 1600s. The amazing drawings in his books laid the foundations of our knowledge of anatomy.

Medical milestones

- 1543: Vesalius published his book *De Humani Corporis Fabrica*—which means "what the human body is made of."
- 1550: Gabriel Fallopio studied the body in minute detail.
- 1590: Santorio Sanctorius created the science of physiology, and showed how to measure pulse and temperature.
- 1628: William Harvey showed how the heart circulates blood.
- 1661: Marcello Malphigi saw tiny blood vessels called capillaries under a microscope.
- 1798: Edward Jenner showed people could be vaccinated against infectious diseases.
- 1840s: Rudolf Virchow and Jakob Henle showed that the body is built up from tissues made from tiny cells.
- 1900: Karl Landsteiner showed that people's blood belongs to different groups.

19

MONITORING THE BODY

A great deal can be learned about the body even without looking inside with X-rays and scanners. The outside of your body gives, surprisingly, many clues as to what is going on inside.

A doctor might learn, for instance, that the airways in your lungs are blocked in a certain way by listening to the sounds of your chest with a simple listening device called a stethoscope. The pattern of pulses that can be felt in blood vessels on your wrist (see p50) might show your heart is unhealthy. Measuring the temperature beneath your tongue might indicate your body is finally overcoming a disease. Even simple visual clues such as inflammation can be surprisingly revealing.

RUNNING ON
Athletes in training want to achieve peak fitness. So many professionals have their own personal physiologist to keep a check on their state of health. The performance of their cardiovascular system—the breathing and circulation—is crucial. Many athletes strap blood pressure and heart pulse monitors on their bodies to tell them what is going on at every moment—and so adjust their training program.

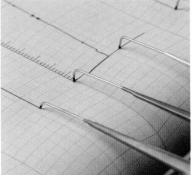

ELECTRO-CARDIOGRAM
An ECG shows the electrical activity of the heart and how the muscles that make the heart beat are working.

CHECKUP
Health checks include measuring blood pressure (see panel) and listening to your chest with a stethoscope for breathing abnormalities.

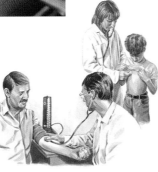

Measuring blood pressure

- Blood pressure is measured with a device called a sphygmomanometer. This has an inflatable rubber cuff that goes around your arm. Attached to the cuff is a pressure gauge.
- The doctor places the cuff around your arm and pumps it up by hand until it puts pressure on the arteries.
- The force with which the blood pushes back against the inflated cuff is shown on the dial of the sphygmomanometer— usually in millimeters of mercury (mm Hg).
- Two pressure readings are taken: the higher "systolic" pressure when the heart is pumping; and the lower "diastolic" pressure when the heart is relaxed.
- To work out when the pressure is systolic and when diastolic, the doctor listens to the sounds of the blood flow with a stethoscope.

Besides the more obvious mechanical tests, doctors can attach electrodes to detect the pattern of electrical activity of things such as the heart (ECG) and the brain (EEG). Samples of blood and urine can also be sent off to laboratories for chemical tests and examination under a microscope.

LOOKING INSIDE

X-rays provide a simple and effective way of taking photographs of the inside of the body, and doctors have come to rely on them to diagnose problems such as chest diseases and bone fractures. X-rays are, like light, a form of electromagnetic radiation. But their waves are very energetic—so energetic that they actually pass clean through some body tissues, as a flashlight shines through thin paper. X-rays are invisible because their waves are too short for our eyes to see, but they register on photographs. By shining X-rays through the body onto photographic film, doctors can take pictures of the inside of the body. The X-rays pass through some tissues and turn the film black—but they are blocked by dense tissues, leaving white shadows on the film.

X-RAY
For an X-ray picture a patient usually lies on or stands against the photographic plate. The radiographer then lines up the X-ray tube to give the clearest possible view through the body. Since an X-ray photograph is not quite as quick as an ordinary photo, the radiographer asks the patient to hold his breath and keep as still as possible while the exposure is made.

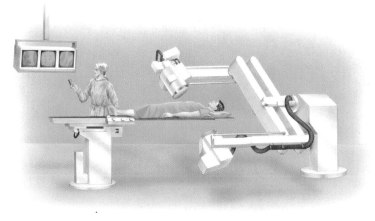

SHADOW PICTURES
Bones produce strong white shadows
in these X-rays of a hand (right)
and the chest (above). X-rays shine
straight through skin and muscle, however,
creating dark areas on the film.

SEE THROUGH TISSUES

Each of the body's tissues absorb X-rays in a certain way. Bones are dense and contain calcium, so they block X-rays almost altogether. They will always show up as white shadows on an X-ray. On the other hand, soft tissues such as skin, fat, blood, and muscle are almost transparent for X-rays, and so go black.

X-rays
How X-rays are made

- X-rays are made by firing streams of electrons at a target of heavy tungsten metal. The beams that bounce back off the target are X-rays.

- X-ray radiation is dangerous in high doses, so the beam is encased in lead, and the radiographer stands behind a screen while the picture is being taken.

How X-rays are used

- X-rays are very good for showing up bone and joint problems. If you break a bone, the doctors will probably take an X-ray to examine the break properly.

- X-rays are also good for analyzing chest and heart problems. If you suffer from chest problems such as TB, pneumonia, or even cancer, the illness may produce dense white shadows on the X-ray where airways in your lungs are blocked.

23

3D PICTURES

In the last 20 years or so, scientists have developed all kinds of clever ways of "imaging" (taking pictures of) the inside of the body, including MRI scans, PET scans, and CAT scans. CAT or Computerizd Tomography scans use X-rays but make the most of computer technology to give doctors incredibly detailed slices through the body.

The X-ray gun rotates around the porthole firing a series of low-dose X-ray beams

Opposite on the porthole to the X-ray gun is a pickup that rotates with it

CAT SCANS
To make a CAT scan, a patient slides into a porthole in the scanner. Once in place, a huge series of X-ray beams is fired across the porthole from numerous different angles. Light detectors on the opposite side of the hole pick up the rays. A computer analyzes what happened to each of the rays as it passed through the patient's body and uses the result to build a detailed picture.

By doing a series of slices in different places, doctors can build up a 3D picture

The CAT scan gives a cross-section slice through the body

BRAIN SCANS

CAT scans are especially good at showing deep inside the brain and revolutionized the treatment of tumors and head injuries.

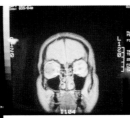

CAT scan showing a slice across the top of the head

CAT scan showing a slice down through the head

The patient lies on a sliding table that moves her in and out of the scanner at a controllable rate

VIEWING THE SCAN

When doctors get CAT scans, they do not get just one picture as with conventional X-rays. Instead they get a whole series of pictures showing slices through the body at very slightly different places. This helps them pinpoint small defects and build up a complete 3D picture.

Scanning techniques

MRI

- MRI or Magnetic Resonance Imaging works a bit like CAT scans, but uses magnets, not dangerous X-rays, and gives much clearer pictures.

- For an MRI scan, the patient is surrounded by a magnetic field so powerful that all the protons (tiny atomic particles) in the body line up in the same direction, instead of all directions as normal.

- A pulse of radio waves knocks the protons briefly out of alignment. As they snap back into alignment a split second later, they send out radio signals themselves. The scanner picks up all the billions of little radio signals to give its picture.

PET scan

- PET (Positron Emission Tomography) scans involve injecting a patient with slightly radioactive substances. Once in the blood, these emit positrons (positively charged electrons), which send out rays which the scanner picks up.

UNDER THE MICROSCOPE

The power of the microscope has revealed a huge number of things about the human body too small to see with the naked eye. It was the microscope that revealed to scientists that the body is made up of more than 60 million tiny cells, and that there are many different kinds. As microscopes have become more and more powerful, they have revealed more and more about the inner workings of the body cells, right down to the level of individual chemical molecules. Microscopes, too, have revealed such tiny structures as the sperm and egg from which every human begins, antibodies in the blood that fight germs, blood cells, and much more.

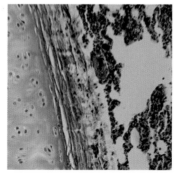

LUNG TISSUE
It needs a microscope capable of magnifying things more than 2,000 times to reveal the fine structure of tissues that make up the lungs. In this picture we see on the left, one of the tiny blood vessels that channel the blood close to the thin walls through which oxygen seeps from the air sacs of the lungs.

Kinds of microscope

- The simplest microscopes use magnifying glass lenses and a beam of light. This is called light microscopy, and can magnify things up to 1,500 times and so reveal individual blood cells.
- To see things even bigger, scientists use beams of electrons, not light, in scanning electron microscopy (SEM) or transmission electron microscopy (TEM).
- SEM can magnify things up to 100,000 times to show such things as the structure of microorganisms.
- TEM can magnify things up to five million times to show such things as the minute individual structures in the cells in a slice of skin.
- To prepare a sample for light microscopy, it must be sliced thinly and placed between two thin sheets of glass, called slides.
- To prepare a sample for SEM, it must be coated with a thin film of gold.

SPOTTING DISEASE

Microscopes have not only revealed a great deal about how the body works, but play an important part in diagnosing illness. By studying blood samples under a microscope, doctors may be able to see germs or parasites, or see if there is anything unusual about the blood cells, as there is when people are anemic.

NERVE CELLS

The amazing spidery nerve cells that make up our brains were only discovered when, in 1873, Italian scientist Camillo Golgi accidentally knocked a piece of owl's brain into silver nitrate. This stained the cells and made them visible under a microscope.

WHAT THE BODY'S MADE OF

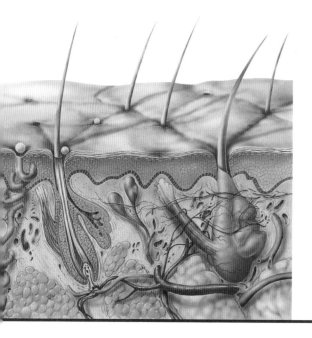

Your body is made of the same basic materials as every living creature—mostly water, proteins, and minerals. But it is the remarkable way these materials are put together that makes the human body so astonishing.

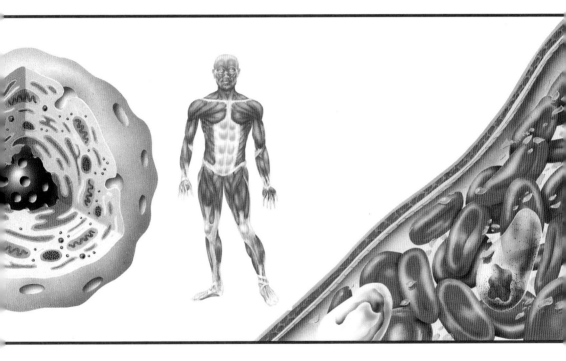

BODY CELLS

Cells are the basic building blocks of our bodies. Just as a house is built of bricks, so bodies are built of microscopic cells. There are 60 million or so in the body, each with its own special task. Some are skin cells. Some are liver cells. Some are fat cells. Each system in the body has its own special cells.

Body cells come in many shapes and sizes, but they are all squidgy cases of chemicals. Holding them together is a thin casing of fat dotted with protein called the membrane. Although it holds the cell together, it lets certain chemicals move in and out. Inside the cell is watery fluid called cytoplasm, and floating inside the cytoplasm is the cell's nucleus—the control center that contains the chemical instructions for all the cell's tasks. Every time a new chemical is needed, the nucleus sends the instructions to the rest of the cell.

LIVING CHEMICAL FACTORY

Every cell in your body is a bustling chemical factory. Inside, every second of the day, the cell's team of "organelles" is ferrying chemicals to and fro, breaking up unwanted chemicals, making new ones, using them and sending them off to other cells. Instructions come from the nucleus, but every bit of the cell has its own allotted task.

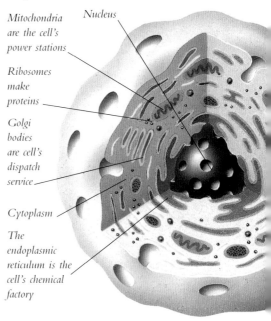

Mitochondria are the cell's power stations

Ribosomes make proteins

Golgi bodies are cell's dispatch service

Nucleus

Cytoplasm

The endoplasmic reticulum is the cell's chemical factory

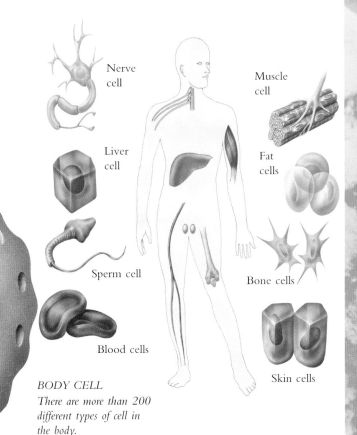

Nerve
cell

Muscle
cell

Liver
cell

Fat
cells

Sperm cell

Bone cells

Blood cells

Skin cells

BODY CELL
There are more than 200
different types of cell in
the body.

Nucleus and organelles

- The nucleus is the cell's
  control center, housing the
  remarkable spiral-shaped
  chemical molecule DNA.
  In a chemical code, DNA
  not only gives all the
  instructions the cell ever
  needs to carry out its task,
  but also carries a complete
  blueprint for another you.
- Outside this is a mixture of
  dissolved chemicals and
  floating structures called
  organelles. Each organelle
  has a particular function.
- Mitochondria are the cell's
  power houses, turning
  chemical fuel supplied by
  the blood into energy packs
  of the chemical ATP.
- The intricate layers of the
  Golgi bodies are the cell's
  despatch center, where
  chemicals are stowed in
  tiny membrane bags.
- Ribosomes make proteins
  from basic chemicals called
  amino acids.
- Lysosomes are the waste
  disposal units, breaking up
  unwanted material.

TISSUES

 All the body's many kinds of cells group together to make substances called tissues. Each tissue is made almost entirely of particular kinds of cells packed together, and each has its own purpose. Nerve tissue, for instance, is made of identical cells called neurons, which are good at sending electrical signals. Epithelial tissue, which makes skin and other tissues, is made of three kinds of cells that make a waterproof layer. Typically, the spaces between tissue cells are filled with a fluid called tissue fluid.

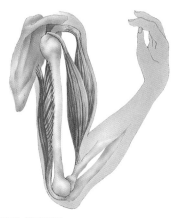

MUSCLE TISSUE
Muscle tissue makes up the bulk of your body—almost half its entire weight. It is tissue built of special long, usually reddish cells that are able to pull shorter and then relax.

EPITHELIAL TISSUE
Epithelial tissue is lining tissue, built up from three basic cell shapes—squamous (flat), cuboid (boxlike), and columnar (rodlike). Most is in layers just one cell thick—lining inside blood vessels, airways, inside the heart and chest and many other places. But skin is made of many layers.

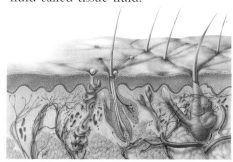

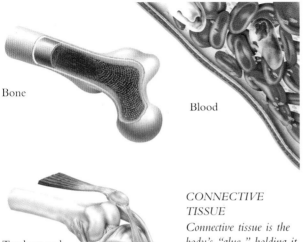

Bone

Blood

Tendons and cartilage on the knee joint

CONNECTIVE TISSUE

Connective tissue is the body's "glue," holding it together, but it comes in a huge variety of different forms.

CONNECTIVE TISSUE

"Connective" tissue forms in the space between other tissues and helps hold the body together. But it comes in a variety of different forms including "adipose tissue," better known as fat, tendons, and cartilage. Bone and even blood are connective tissues. Of course, blood does not actually connect anything, but it begins life in the human embryo in the same way.

Tissues

Organs

- Different types of tissue group together to make organs.
- The bulk of your heart is made of muscle tissue. But it is lined with epithelial tissue, nerve tissue sends signals to it, and it is filled with the connective tissue of blood.

Connective tissue

- Connective tissue is made of three main things: cells, fibers, and "matrix."
- Cells are mainly "fibroblasts," cells that make fibers. But they also include fat cells and "macrophages," which eat germs.
- The fibers are made from microscopic strings of protein such as white ropey collagen, stretchy elastin, and branchlike reticulin.
- The matrix is the basic setting for the other materials, like the bread in a raisin loaf. It can be anything from a runny syrup to a thick gel.

33

BODY SYSTEMS

Although your body seems very complicated, it all makes sense if you think of it in terms of a number of systems, each with its own tasks to do. It only seems complicated because these systems are all interlinked.

Some of these systems extend throughout the body, such as the skeleton, which is the body's framework; the musculature, which is the body's means of moving; and the nervous system, which is the body's communication networks. Others are

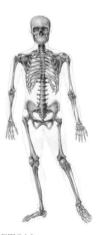

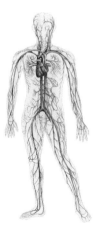

SKELETON
The skeleton supports the body and protects internal organs such as the lungs and the heart.

MUSCULATURE
The muscles are what enable you to move. Muscles are also involved in other body systems.

CARDIOVASCULAR SYSTEM
Heart and blood circulation supply body cells with oxygen and food, and defend the body against germs.

NERVOUS SYSTEM
The nervous system is the brain and the nerves—the body's high speed control network.

quite localized, such as the digestive system, which is the body's food processor, and the urinary system, which controls the body's water. The body could not function without any of these systems, and you go on living because they all work together to keep you alive.

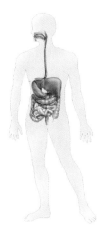

DIGESTIVE SYSTEM
The digestive system breaks down food and turns it into the right chemicals for the body to use.

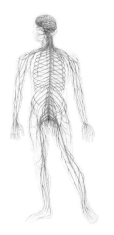

LYMPHATIC SYSTEM
Lymph fluid containing immune cells that help fight disease circulate in lymphatic ducts.

Body systems
- Skeleton, or skeletal system.
- Musculature, or muscle system.
- Cardiovascular system.
- Nervous system.
- Digestive system.
- Excretory system.
- Urinary system: controls the body's water and removes excess as urine.
- Immune system, including the lymphatic system, white blood cells, and antibodies: the body's various defenses against germs.
- Respiratory system: takes air in and out of your lungs through the mouth and nose, giving oxygen to the blood and taking out carbon dioxide.
- Reproductive system: the smallest of the systems, this is essentially the genitals and the organs connected to them—the organs that enable us to have children. This is also the only body system that can be surgically removed without threatening your life.

35

BREATHING

Every tiny cell in your body needs an almost continuous supply of oxygen, just to survive. That is why you can never stop breathing in air, for even a minute—and why your body has developed an amazing system for extracting huge amounts of oxygen from the air every few seconds.

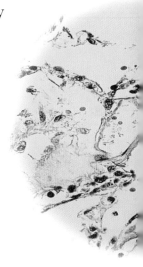

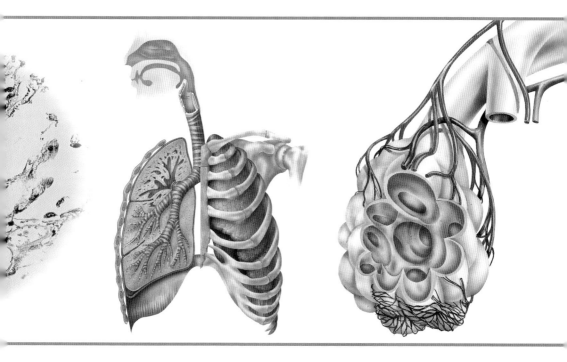

BREATHING

You have to breathe to stay alive. When you breathe in, you take air into the body—and air contains the oxygen vital to every cell in your body. Just as a fire needs air to burn, so body cells need oxygen to burn up the food they get in the blood. Without it, they die—which is why if you stopped breathing for even a few minutes, you would lose consciousness, and eventually die. So, several times a minute your body sucks air into your lungs so that oxygen can be spread around the body in the blood.

BREATHING AT ALTITUDE

The air high on a mountain contains less oxygen, so mountaineers must breathe and circulate blood faster to compensate—or carry a supply of oxygen.

PANTING HARD

The cells in a runner's muscles are working overtime, burning up sugar at a huge rate to keep him pounding along. Burning all this sugar means the cells need lots of oxygen, so he must breathe hard and his heart must pump blood fast to supply the muscles with all the oxygen they need. As a runner trains and gets in shape, his body gets much quicker at raising oxygen levels. The out of shape are left behind, gasping for breath.

WHY BREATHE OUT?

Food arrives at each body cell as a chemical called glucose, which is made mainly of carbon and hydrogen. When the cell burns glucose to release its store of energy, the hydrogen joins with oxygen to make water and the carbon joins with oxygen to make carbon dioxide. Carbon dioxide is poisonous to the body and must be removed. This is what happens when you breathe out.

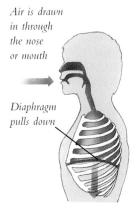

Air is drawn in through the nose or mouth

Diaphragm pulls down

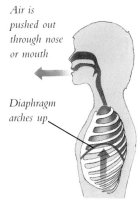

Air is pushed out through nose or mouth

Diaphragm arches up

BREATHING IN

To draw air in, the lungs must expand. "Intercostal" muscles around your ribs pull your chest up and out, and a domed sheet of muscle called the diaphragm pulls the lungs down.

BREATHING OUT

Lungs are spongy and elastic, so collapse like a deflated balloon as you breathe out—but can only collapse so far because a layer of "pleural fluid" sticks them to the chest walls.

Breathing

- You will probably take about 600 million breaths if you live to the age of 75.
- Every minute you breathe, you take in about 14 pints (6 l) of air.
- A normal breath takes in about a pint (0.4 l) of air. A deep breath can take up to 10 pints (4 l).
- On average, you breathe about 13-17 times a minute. But if you run hard, you may have to breathe up to 80 times a minute.
- Newborn babies breathe about 40 times a minute.
- Breathing is called respiration. The way cells burn up glucose using oxygen is called cellular respiration.
- Air normally contains about 21 percent oxygen, and about 0.03 percent carbon dioxide. The air you breathe out has about 0.6 percent carbon dioxide—not quite enough to poison anyone with your breath!

THE LUNGS

Your lungs are a pair of soft bags in your chest. Like spongy cushions, they are riddled with tiny branching tubes that join together to lead right up to your mouth. Whenever you breathe in, air is sucked in through your mouth or nose and rushes down your windpipe or trachea until it reaches a fork deep inside your chest. At the fork, the airways branch into two, one branch or bronchus leading to the left lung and the other to the right.

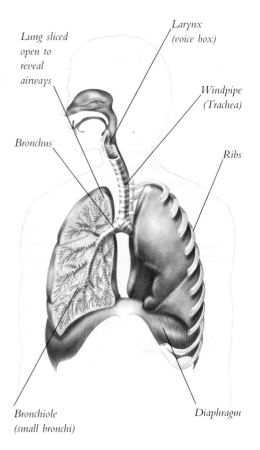

Lung sliced open to reveal airways

Larynx (voice box)

Windpipe (Trachea)

Bronchus

Ribs

Bronchiole (small bronchi)

Diaphragm

INSIDE THE LUNGS

Inside the lungs, the bronchi branch into millions of bronchioles. There are so many tiny airways in the lungs that they have a gigantic surface area. It is this huge surface area that allows a great deal of oxygen to seep through into the surrounding blood vessels in just a few seconds every time you breathe.

ALVEOLUS

Alveoli (right) are the tiny air sacs at the end of each airway that fill up like balloons as you breathe in. Here oxygen passes through into the blood.

LUNG TISSUE

Under a microscope, you see that lung tissue (right) is filled with tiny blood vessels, making it easy for oxygen to pass quickly through into the blood.

Capillaries

Lung tissue

AIR TO BLOOD

It is in the lungs that oxygen in the air is absorbed into the blood. Around the end of each bronchiole are bunches of tiny air sacs called alveoli. The walls of each sac are just one cell thick, and here oxygen seeps through into the minute blood vessels wrapped around them.

Lungs

The lung surface

- There are about 300 million alveoli in your lungs.
- Laid flat, the inside of the alveoli would cover an area the size of a tennis court.
- There are more than 1,500 miles (2,400km) of airways in your lung.

Coughing and hiccupping

- The surfaces of the airways are protected by a film of slimy liquid called mucus. When you have a cold or a chest infection, the airways may fill up with mucus, making you cough to clear them.
- Smoking irritates the airways and makes them fill up with mucus. It also weakens the tiny hairs or "cilia" that waft the mucus out. So lungs are prone to infection. Smoking also increases the risk of lung cancer.
- Hiccups are caused by a sudden contraction of the diaphragm, dragging air into your lungs so quickly that your vocal cords snap shut.

41

SPEAKING AND SINGING

Speaking and singing are so natural that it is hard to think about just how you do it. The secret lies in your throat. This is where the larynx or voice box is. The voice box is not actually a box at all. It is just the part of the airway between your lungs and your mouth where the vocal cords are located. The vocal cords are bands of muscle that vibrate when air whistles past them as you breathe out. Most of the time the vocal cords leave an open hole for air to pass through as you breathe. But when you speak or sing, the cords tighten across the throat and vibrate enough to produce sounds. By widening or narrowing the gap between the cords, the muscles of the larynx can vary the pitch, to make a high or low sound.

TRIBAL SOUND
Many African peoples have developed extraordinary ways of controlling the sounds produced by their vocal cords. Masai warriors, for instance, make the most of all the echoes they can make with the bones of their head to create a very loud chanting sound. When just a few Masai warriors make this mouth music together, the sound can be heard over huge distances.

Vocal cords
from above

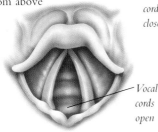

*Vocal
cords
closed*

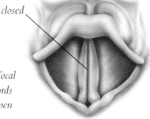

*Vocal
cords
open*

LOW SOUND
*When the vocal cords are
wide open, the cords
vibrate slowly, creating a
low-pitch sound.*

HIGH SOUND
*When the vocal cords are
close together, they
vibrate rapidly to create a
high-pitch sound.*

SPEAKING
*The vocal cords give
a simple "aaah"
sound. But by
changing the shape
of your pharynx
(upper throat), your
lips, mouth, and
tongue, you can
change the sounds
emitted by the vocal
cords into words.*

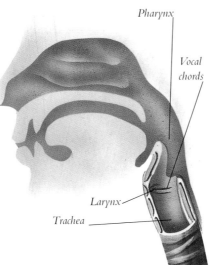

Pharynx

*Vocal
chords*

Larynx

Trachea

Voice

- When babies are born, their
  vocal cords are about 1/4
  inch (6mm) across.
- A man's vocal cords are
  about 1 1/4 inches (30mm)
  long; a woman's are about
  3/4 inch (20mm). Because
  a man's vocal cords are
  longer, they vibrate more
  slowly and give him a
  deeper voice.
- During childhood, a boy's
  vocal cords are similar in
  length to a girl's. But in his
  teenage years, they grow
  longer, and his voice
  "breaks," becoming deeper.
- At the front, the vocal cords
  are attached to a small
  lump of cartilage (see p73)
  called the Adam's apple. In
  his teenage years, a boy's
  Adam's apple grows much
  bigger and is usually visible
  at the front of the neck.
- The technical name for the
  vocal cords and the slit
  between them is the glottis.
  The epiglottis is the fold of
  cartilage above the glottis
  that blocks off the larynx
  when you swallow food.

43

HEART AND BLOOD

Your heart is the most amazing little pump in the world, tirelessly beating away inside your chest every second of your life to keep blood circulating around your body. Every cell in the body depends entirely on the array of materials brought to it in the blood that the heart delivers— and without it the body would quickly die.

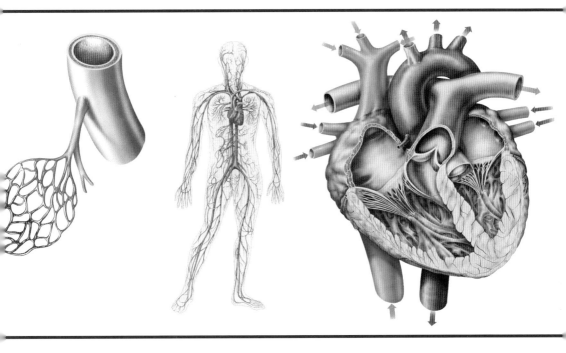

CIRCULATION

Once in the blood, oxygen must be delivered swiftly to the cells where it is needed. At the same time, unwanted carbon dioxide must be picked up from the cells and brought back to the lungs for breathing out. This is what blood circulation is for.

Driven by the heart, blood is pumped through an intricate network of blood vessels all the way around the body again and again, once every 90 seconds. As it circulates it goes through the lungs where it washes around the lung's millions of air sacs and picks up oxygen. This bright red, oxygen-rich blood then goes back through the heart, where the heart's powerful right side gives an extra boost to push it into

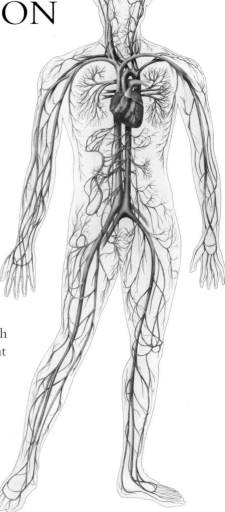

BLOOD VESSELS
The blood circulates through the body in a series of pipes called blood vessels. Oxygen-rich blood is bright red and travels through arteries on the way out from the heart. Oxygen-poor blood is purple and travels in veins on its way back to the heart.

46

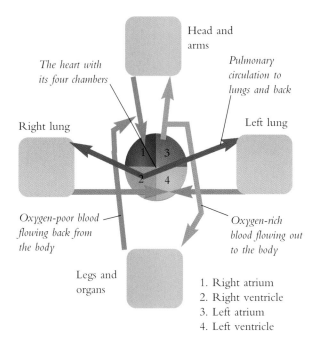

The heart with its four chambers

Head and arms

Pulmonary circulation to lungs and back

Right lung

Left lung

Oxygen-poor blood flowing back from the body

Oxygen-rich blood flowing out to the body

Legs and organs

1. Right atrium
2. Right ventricle
3. Left atrium
4. Left ventricle

the arteries and right around the body. As it passes into ever narrower blood vessels—first arterioles, then capillaries—it delivers its oxygen to the cells and picks up the waste carbon dioxide. It then flows back through the veins to the left side of the heart—now much bluer in color because it is short of oxygen—ready to be pumped on to the lungs for new intake of oxygen.

Circulation

- The blood circulation has two parts: the systemic and the pulmonary.
- The pulmonary circulation is the short section that sends oxygen-poor blood from the heart to the lungs and brings it back to the heart with fresh oxygen.
- The systemic circulation sends oxygen-rich blood on from the right side of the heart and carries it right around the body and back to the left side of the heart.
- Oxygen is ferried through the blood in red blood cells.
- Each of the millions of red blood cells can carry plenty of oxygen because they contain a substance called hemoglobin, which takes up oxygen when there is plenty around, and gives it up when there is a shortage.
- Hemoglobin glows bright scarlet when it is carrying oxygen, which is why blood is red. But it fades to dull purple when it loses oxygen.

THE HEART

The heart is a remarkable little pump, about the size of a fist and made of pure muscle. Every second of your life, the heart muscle is squeezing away, pumping blood around your body in a continuous stream. About 70 times a minute—faster if you're running around—the muscles contract to send jets of blood shooting through the arteries, the big pipes that lead away from the heart. Press a hand against the middle of your chest, just to the left, and you can often feel the heart pounding away inside.

The heart owes its steady beat to the special muscle it is made from. Most muscle has to be prodded into action by nerve signals. Heart muscle contracts and relaxes rhythmically by itself. If it has blood to keep it going, it can go on working even outside the body. Outside stimuli can only change its rate.

Superior vena cava—main vein bringing blood from the body

Pulmonary artery taking oxygen-poor blood to lungs

Pulmonary vein bringing oxygen-rich blood from the lungs

Right atrium

Tricuspid valve between right atrium and right ventricle

Right ventricle

INSIDE THE HEART

The heart is not just one pump but two, each forming one side of the heart. A thick wall of muscle called the septum keeps the sides completely separate. The left side is the stronger pump, driving oxygen-rich blood from the lungs all around the body. The right is the weaker pump, pushing it only through the lungs to pick up oxygen. Each side has two chambers separated by a one-way valve—an "atrium" at the top, where blood accumulates, and a "ventricle" below, which is the main pumping chamber.

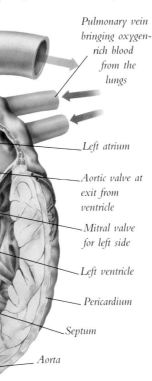

Aorta—main artery taking blood to body

Pulmonary artery taking oxygen-poor blood to the lungs

Pulmonary vein bringing oxygen-rich blood from the lungs

Left atrium

Aortic valve at exit from ventricle

Mitral valve for left side

Left ventricle

Pericardium

Septum

Aorta

Inferior vena cava—main vein bringing oxygen-poor blood from the body

Heart

- The heart beats more than 30 million times a year.
- All the chambers of the heart hold about the same quantity of blood—about 3 fluid ounces (70-80ml).
- The heart has four valves to ensure blood flows only one way, two on each side of the heart—a large valve between the atrium and ventricle and a smaller one at the exit to the ventricle.
- The large valves have different names on each side of the heart. The valve in the left is called the mitral valve; the valve in the right is called the tricuspid valve.
- Heart muscle gets its blood from the blood vessels supplied by the coronary arteries, which wrap around the outside. If these arteries get blocked, the heart muscle may be deprived of blood and stop working. This is what happens in a heart attack.

49

BEATING HEART

Much of the time, you would hardly know your heart is there. But there are always faint signs of its tireless beating. Every second or so, a tiny shock wave or pulse runs through the blood and can be felt in places where major arteries near the body surface. This shock wave is set off by the snapping shut of the valves which are the key to the heart's pumping action. There are four of them altogether, one at the exit of each chamber of the heart.

Each of these valves is a trap door that swings open one way only. As each ventricle empties, blood in the atrium pushes the valve between them open so that blood floods into the

HEART CYCLE

Every time the heart beats, it goes through the same sequence of events, called the cardiac cycle. This has two phases: systole (contraction) and diastole (relaxation). Systole begins as a wave of muscle contraction runs down across the atria from left to right, driving blood into each of the ventricles. A fraction of a second later, the wave reaches the ventricles, and they too contract, squeezing blood into the arteries. Then in diastole, both atria and ventricles relax, the atria fills with blood and the cycle starts again.

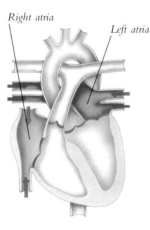

Right atria

Left atria

1. Blood floods into the relaxed atria

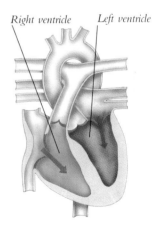

Right ventricle   Left ventricle

2. The wave of contraction squeezes the atria, driving blood into the ventricles

ventricle. But as the ventricle fills, the pressure of blood pushes the valve shut again. When the heart muscles contract to squeeze the ventricle, the blood pushes open the valve at the ventricle's exit—so the blood gushes down the artery away from the heart.

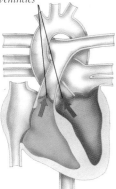

Semilunar valves at exit of ventricles

3. *Blood squeezed in the ventricles pushes open the semilunar valves*

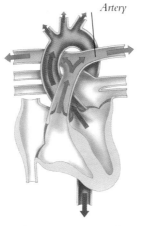

Artery

4. *The ongoing contraction drives blood into the arteries*

Heart beat

- Doctors check the pulse rate for clues to health.
- Pulse rates vary between 60 and 100 beats a minute. About 75 is average.
- Vigorous exercise, or a shock, can push the heart rate up dramatically.
- A heart rate of more than 100 beats a minute is called tachycardia.
- If someone has tachycardia when sitting down normally, it may be caused by too much coffee or tea, drugs, anxiety, or fever—or may be due to a heart problem such as coronary artery disease.
- A heartbeat slower than 60 beats a minute is called bradycardia.
- Any abnormality in the heartbeat is called arrhythmia. If doctors suspect a problem, they connect the patient to an ECG, or electrocardiogram.
- The wave of contractions in heart muscle depends on electrical impulses.

51

ARTERIES AND VEINS

Threading through the body are literally millions of blood vessels—some as wide as a pen, some as thin as a hair. The biggest blood vessels carrying blood away from the heart are the arteries. These branch into arterioles, and these in turn branch into capillaries.

From the capillaries, blood flows into wider venules, and then into wider still veins on its way back to the heart. Blood vessels are not just passive pipes. All but capillaries have muscles and valves to control the way blood flows, helping smooth out the flow.

BRANCHING NETWORK

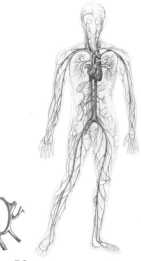

The body's network of blood vessels is like two river systems intertwined. This illustration shows only the arteries and arterioles in red in one half of the body and only the veins and venules in blue in the other. Of course, both veins and arteries supply both sides of the body, and the networks are duplicated.

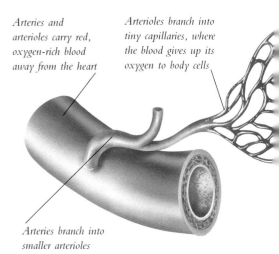

Arteries and arterioles carry red, oxygen-rich blood away from the heart

Arterioles branch into tiny capillaries, where the blood gives up its oxygen to body cells

Arteries branch into smaller arterioles

The middle layer of a vein wall is muscle that contracts or expands to open it or close it.

The tough, fibrous outer layer of a vein wall is elasticated, and has its own blood supply of tiny capillaries

The smooth, inner layer of the vein wall is also elasticated

Venules deliver blood into bigger veins

Once the blood has given up its oxygen, it loses its bright red color and turns purple

BLOOD VESSELS
Arteries and veins are the blood circulation's main highways. They are linked by small arterioles and venules and tiny capillaries.

Blood vessels

- Blood races through the arteries at 3 feet (90cm) a second.
- There are more than 36,000 miles (60,000km) of capillary in the body.
- The muscles of the blood vessel walls control the pressure of blood flow. They tighten and relax so that no matter how much blood is delivered by the main arteries, pressure in the capillaries is always right. It must be strong enough to push oxygen to every cell, but not so strong as to burst the capillaries.
- High blood pressure is called hypertension. It is caused by the thickening of the artery walls, which makes the arteries narrow.
- Blood pressure is lower in veins than it is in arteries.
- The walls of veins are thinner, weaker, and less elastic than those of arteries. They even collapse when empty, unlike arteries.

DYNAMIC PIPES

The muscular walls of the blood vessels control the flow of blood, narrowing or widening them, for instance, to divert blood to where it is needed. When tissues such as muscles are active, blood vessels to them open up to increase blood supply. When tissues are resting, some of the blood vessels close.

BLOOD

Blood is the body's transportation system. It not only carries vital oxygen from the lungs to every cell in the body. It also carries all the food needed to fuel and maintain each cell, and washes away all the waste to the liver, kidney, and lungs for disposal. It helps spread body heat, too. It even plays a crucial role in the body's defense against disease.

WHAT'S BLOOD MADE OF?

Blood looks like red ink, but under a microscope, you can see a rich variety of ingredients, swept along in a clear, yellowish fluid called plasma. Besides many chemicals, there are three main kinds of cells. Most numerous are the button-shaped red blood cells that carry oxygen. Then there are tiny lumps called platelets. Finally there are giant white cells or "leucocytes."

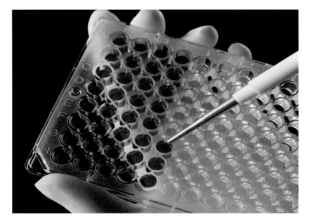

BLOOD SAMPLE
Blood need only be taken in very small quantities to reveal a great deal about it. There are, for instance, five million red blood cells in even a small drop of blood. Red cells, white cells, and platelets make up half the blood in any sample. The rest is the fluid plasma, which includes a solution of the chemical albumin, antibodies (see p202) and "clotting factors"—chemicals that make the blood clot (thicken to stop bleeding).

DROP OF BLOOD

Under a microscope, you can see some of the many different ingredients in blood. It is dominated by the red blood cells which turn bright red only when they are carrying oxygen. This gives blood its red color. Otherwise, blood is purply brown.

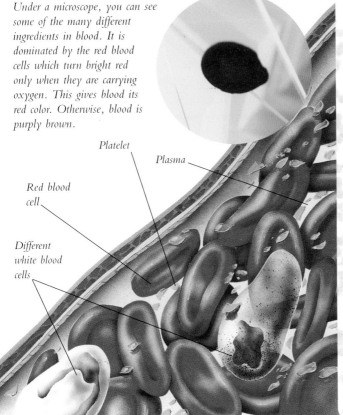

Platelet

Plasma

Red blood cell

Different white blood cells

Blood cells

Red blood cells

- Red blood cells are the button-shaped cells that contain hemoglobin, the special chemical that holds oxygen as it is ferried by the cell around the body.

White blood cells

- White cells are involved in the body's defense.
- Most white cells contain tiny grains and are called granular leucocytes or granulocytes.
- Most granulocytes are called neutrophils, which act as scavengers, eating up intruders (see p200).
- There are two other kinds of granulocyte: eosinophils and basophils.

Platelets

- Platelets are bits chipped off other cells. They help to plug leaks such as cuts.
- Platelets also slow blood loss by releasing "clotting activators," which help fibers grow around the wound and stop blood from leaking.

BLOOD IN ACTION

Blood is not just a still liquid. It is always buzzing with activity. Some cells in the blood are multiplying and changing. Some are actively fighting disease and some are moving here and there dealing with day-to-day emergencies.

Few blood cells live long. Red cells die after four months. Neutrophils survive for six hours or so. So the blood must constantly be topped up with new cells to replace those whose time is up. Nearly all these cells are born, not in the blood, but in the soft red honeycomb called the marrow in the middle of certain bones, such as the breastbone, spine and ribs. All the different cells start life as one kind of cell called stem cells and gain their identities as the cells divide.

PLUGGING A LEAK

When you cut yourself, you bleed from the damaged blood vessel and a few platelets gather at the site to help seal the leak. As they clump together, they release chemicals called clotting factors that draw other platelets. The platelets soon become sticky and begin to join together to form fibers, or "fibrin." This is called coagulation.

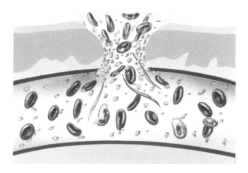

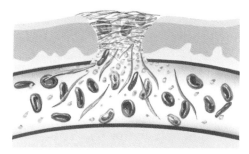

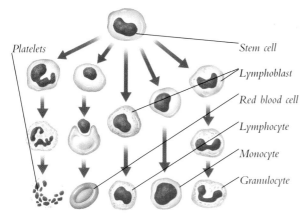

Platelets

Stem cell

Lymphoblast

Red blood cell

Lymphocyte

Monocyte

Granulocyte

HOW BLOOD CELLS FORM
Some stem cells divide to form platelets and red blood cells. Others form lymphoblasts, which divide in turn into white cells—lymphocytes, monocytes, and granulocytes.

FORMING A SCAB
Once the jellylike mass of fibrin has plugged the leak, it dries out and contracts to form a hard scab to protect the wound until it has healed.

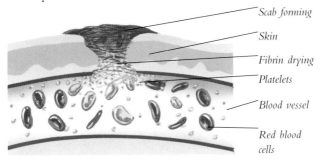

Scab forming

Skin

Fibrin drying

Platelets

Blood vessel

Red blood cells

Blood clotting

- Blood clotting stops leaks from cuts—but if the blood clots too easily, it can clog up arteries.
- A clot that clogs up an artery is called a thrombosis.
- When a clot builds up in one of the arteries of the heart it is called a coronary thrombosis and can cause a heart attack.
- Some people suffer from an illness that stops their bodies from absorbing vitamin K, which the blood needs to make clotting factors.
- Just one molecule of fibrin needs to form at a cut for it to set off a chain of coagulation that creates 30,000 more molecules almost instantly.
- The chemicals involved in clotting are called clotting factors. They are made in the liver and numbered from I to VIII (one to eight).
- Clotting factor I is called fibrinogen and encourages fibrin formation.

57

BLOOD GROUPS

Many people's lives have been saved by receiving blood donated by others. But a blood "transfusion" from the wrong person is not a lifesaver; it can actually be fatal. This is because everyone has blood that belongs to a particular group.

Every group but your own has antigens—that is, identity markers that tell your body it is foreign. These antigens make your body's immune system react against it and can make you very ill.

Most people have blood that belongs to one of four groups: A (the most common), O, B, and AB. These are called the ABO groups, or "factors."

Another set of antigens in blood divides people into Rhesus positive and Rhesus negative, according to whether they have this set of antigens or not. The most important of these antigens is known as factor D.

About 85 percent of people are Rhesus (Rh) positive. If your blood is Rh positive and also group O, your blood group is said to be O positive. There are actually nearly 400 other antigens affecting blood, but these are the important ones.

BLOOD TEST
If you want to give blood, doctors will need to know what blood group you belong to and whether you are a suitable donor. But they need only a little pinprick sample to carry out their tests.

BLOOD BANK
People often donate blood at hospitals to give to people in emergencies. An adult usually has about 11 pints (5 l) of blood, and typically donates 1 pint (0.5 l). The body makes up for this loss completely within a few days. Sometimes, doctors take the whole blood; sometimes just specific components such as plasma or white cells.

Blood products
- After donation, blood is stored either whole or separated into various components. Each has its use in blood transfusion.
- Whole blood is used when people suffer severe blood loss in an accident or operation.
- Red cells are used to treat people who are very anemic—that is, they have a very low red blood cell count.
- Red cells may be stored either washed and separated from other cells or frozen.
- Platelets are given to people who bruise too easily.
- White blood cells called granulocytes (see p55) may be given to people suffering certain dangerous infections.
- Clotting factors help the blood clot and slow bleeding. So factors 8 and 9 are given to people suffering from hemophilia, which makes them bleed too easily.

SKIN AND BONE

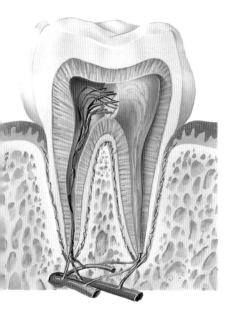

Skin is your body's protective casing and the skeleton its living framework. But they don't just provide protection and support. The skin is the body's largest organ—and also its largest sense receptor by far, providing vital data about the world to your brain. The skeleton is where all the cells for your life-giving blood are born.

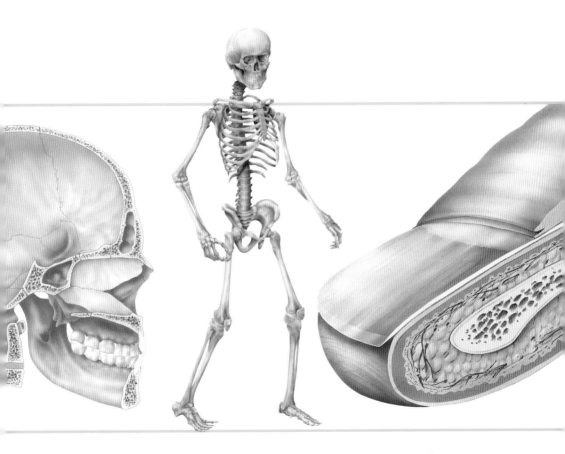

SKIN

Your skin is not just a bag for your body. It is actually the body's largest organ and has a wide range of vital tasks. It is a coat that protects your body from the weather and from infection. It helps keep your body at just the right temperature by insulating it from the cold and letting out heat when it is warm. It helps you sense the world around you by responding to touch. It even helps nourish you, by using sunlight to make vitamin D. It is so important that it receives almost a third of your body's blood supply and has a range of special glands.

SLICE OF SKIN
This is a hugely magnified view of a slice through the skin, showing just some of the elements. Its thickness varies over the body. It is thickest on the soles of your feet, and thinnest on your eyelids, where it is just 1/50 inch (0.5mm) thick.

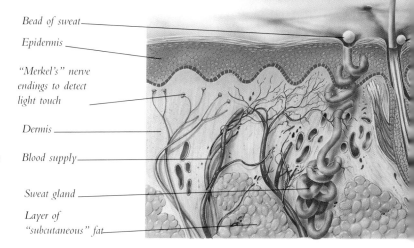

Bead of sweat

Epidermis

"Merkel's" nerve endings to detect light touch

Dermis

Blood supply

Sweat gland

Layer of "subcutaneous" fat

DOUBLE LAYER

Over most of your body, your skin is about 1/12 inch (2mm) or so thick, and is easily penetrated by a sharp edge. A microscope shows it is made of two main layers: a thin, essentially dead, outer layer called the epidermis and a thicker lower layer called the dermis, which contains the glands.

Layer of dead skin

Most areas of the skin have at least small hairs

Hair follicles make an oily substance called sebum that coats your hair and keeps it waterproof

Basal layer of new cells

"Ruffini" nerve endings to detect heat and sustained pressure

Hair erector muscle

Hair follicle

Skin

The epidermis

- New cells are continually growing in the basal layer at the bottom of the epidermis.
- As new cells grow in the epidermis's basal layer, they push old cells up. As they move farther away from the blood supply, they die off, leaving just a hard protein called keratin. The layer of keratin gives the skin a hard, protective coat.
- Dead cells continually flake off the skin's surface, but new cells come up to take their place.

Skin color

- The epidermis contains special cells that produce a dark pigment called melanin. In fair-skinned people, extra melanin is made to protect the skin in strong sunlight, which is why they tan. Dark-skinned people start with much more melanin in their skin. Freckles are just spots of melanin-rich skin.

HAIR AND NAILS

Almost alone among mammals (except for those that live in the sea), we humans have bare skin. Apart from the hair on your head (and around your genitals when you're an adult), your skin has only a light covering of tiny down hairs.

People sometimes talk about dull and lifeless hair when it is in poor condition. In fact, hair is always lifeless, for it is made of keratin, which is the dry material left behind by dead cells—the same dead material that nails and the outer layer of skin is made from.

The root of each hair is embedded in the skin, in a pit called the hair follicle. The hair grows as dead cells pile up within this follicle, held in place by the hair's club-shaped end or "bulb." Each hair has a spongy core, surrounded by a ring of long fibers that are wrapped around in turn by overlapping layers.

DARK HAIR

Providing you don't dye it, the color of your hair depends on just how much of a pigment called melanin you have in the shafts of your hairs. This pigment is made in cells called melanocytes near the hair root.

HAIR BASE

A fingernail takes about six months to grow from base to tip, although it varies with the seasons. Like hair, it is made of keratin and grows from a root, called the nail bed. At the base of each nail is a half moon, called the lunula, covered by a flap of skin called the cuticle.

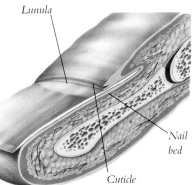

Lunula

Nail bed

Cuticle

HAIR BASE

Hair is made from dead skin cells and keratin and grows in follicles deep in the skin. At the root, there is an erector muscle that pulls it upright when you are cold. There is also a sebaceous (oil) gland that oozes oil to keep the hair waterproof and stop it from drying out.

Hair root.

Sebaceous oil

Hair follicle.

Hair facts

- There are three types of hair on your body.
- Lanugo is the downy hair you were covered with in the womb, from the fourth month to the time you were born.
- Vellus hair is fine, short, downy hair that grows all over your body until you reach the age of puberty.
- Terminal hair is the thicker, longer hair that grows on your head until you reach puberty—then it grows in your pubic region and under your armpits as well.
- Most men also grow terminal hair on their chins.
- The curliness of your hair depends on the shape of the follicle.

Hair color
- Hair is red or auburn if the follicles contain the red kind of the melanin pigment. All other hair colors come from black melanin.

TEETH

Biting into food and grinding it into small and soft enough lumps for you to swallow places enormous wear on teeth. So their outer coating is made of the body's hardest substance, the white material, enamel. Inside the enamel is a softer material called dentine, but even dentine is as hard as the hardest bone. Yet even though they are tough, teeth can be eaten away by the bacteria that thrive in the mouth when you eat starchy and sugary foods. This is why it is so important to clean your teeth well, especially after eating sugary foods. Fluoride is often added to drinking water at the waterworks to protect teeth from decay. Even so, many children and adults must have holes filled by the dentist from time to time.

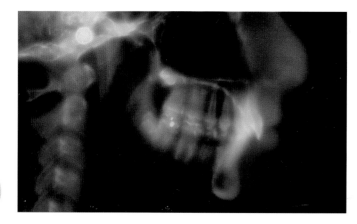

X-RAY OF ADULT JAW
Everyone has two sets of teeth during their lives. The first set of 20 are called baby teeth, and start to appear when you're about the age of six months—a process called teething. When you're about six, the baby teeth begin to be replaced by 32 adult teeth, which must last you all your life. In this X-ray, the tooth roots can be seen clearly.

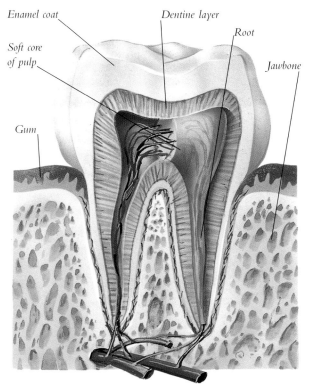

Enamel coat

Dentine layer

Root

Soft core of pulp

Jawbone

Gum

INSIDE A TOOTH

Teeth have long roots that fit into sockets in the jaw bones but they are surrounded by soft gums. At the heart of each tooth is a living pulp of blood and nerves. Around this is a layer of dentine. Around this, where it projects from the gum, is a layer of hard white enamel.

Kinds of teeth

Molars

- Molars are the big strong teeth at the back of your mouth on either side. They have flat tops with ridges and are good for grinding and softening food. You have two or three pairs of molars on each side.

- The third pair of molars, called wisdom teeth, are the last to grow, and in some people, never grow at all.

Premolars

- As an adult, you should have two pairs of premolars on each side. They have two edges but like molars they are good for grinding.

Canines

- You have just one pair of canines on each side. They are the big, pointed teeth just behind your front teeth—ideal for tearing food.

Incisors

- Incisors are the flat front teeth, with sharp edges for cutting food. There are two pairs on each side of the mouth.

THE SKELETON

Your body is held up by a strong framework of bones called the skeleton. The skeleton not only provides an anchor for the muscles, but supports the skin and other tissues, and protects your heart, brain, and other organs. It is made up of more than 200 rigid bones, joined together by softer, rubbery cartilage. The skeleton is the one part of the body that survives after we die, but while we are alive, it is a living tissue and is constantly being renewed as old bone cells die and new ones are born in the bone's core, or marrow.

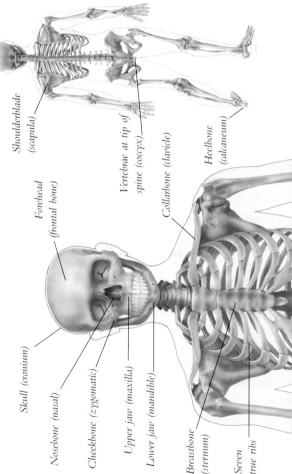

Shoulderblade (scapula)

Vertebrae at tip of spine (coccyx)

Collarbone (clavicle)

Heelbone (calcaneum)

Forehead (frontal bone)

Skull (cranium)

Nosebone (nasal)

Cheekbone (zygomatic)

Upper jaw (maxilla)

Lower jaw (mandible)

Breastbone (sternum)

Seven true ribs

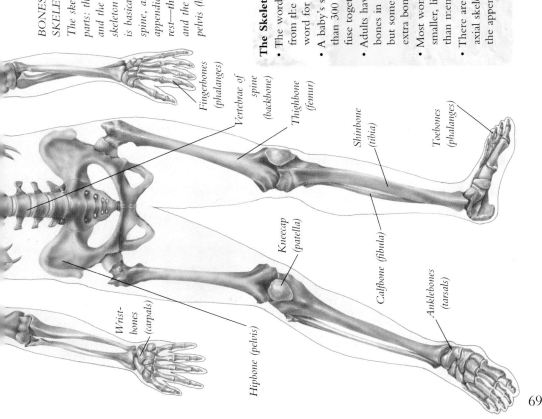

BONES OF THE SKELETON

The skeleton has two main parts: the "axial" skeleton and the "appendicular" skeleton. The axial skeleton is basically the skull, the spine, and the ribcage. The appendicular skeleton is the rest—the arms and shoulders, and the legs and pelvis (hipbone).

The Skeleton

- The word skeleton comes from the Ancient Greek word for dry.
- A baby's skeleton has more than 300 bones, but these fuse together as they grow.
- Adults have on average 213 bones in their skeletons, but some people have extra bones in the spine.
- Most women and girls have smaller, lighter skeletons than men and boys.
- There are 87 bones in the axial skeleton and 126 in the appendicular.

Fingerbones
(phalanges)

Vertebrae of
spine (backbone)

Thighbone
(femur)

Shinbone
(tibia)

Toebones
(phalanges)

Kneecap
(patella)

Calfbone (fibula)

Anklebones
(tarsals)

Wrist-
bones
(carpals)

Hipbone (pelvis)

69

BONES

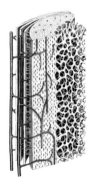

Bones provide our bodies with a strong but remarkably light framework. Their strength comes from their special combination of flexibility and stiffness. The flexibility comes from tough, ropelike fibers of a material called collagen. The stiffness comes from the hard deposits of minerals, especially calcium and

phosphate, around the collagen. Many bones are very light because much of their core is made up from a honeycomb of thin struts called trabeculae. Trabeculae are thin, but perfectly angled to resist stresses.

INSIDE THE BONE
Bones have a surprisingly complex structure. On the outside is a casing of "compact" bone, made up of tiny tubes called osteons. Inside this is a layer of spongy or "cancellous" bone made of a honeycomb structure of trabeculae. Inside this is the soft marrow, where blood cells are made. Covering it all is a membrane, or periosteum.

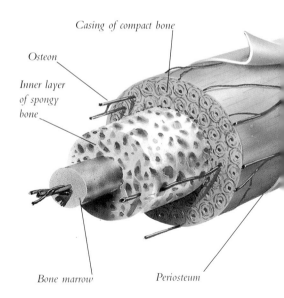

Casing of compact bone

Osteon

Inner layer of spongy bone

Bone marrow

Periosteum

As children grow, long bones get longer at the end, called the epiphyseal plate

Marrow in long bones turns yellow as you get older

BONE MARROW (above) In the hollow centre of the breastbone, ribs, and hips is soft, spongy, red "marrow" where red and white blood cells are created.

INSIDE A BONE (left) Long bones of the arms and legs have a hard outside but a soft inside.

LIVING BONES

Bones are as alive as any of our other body tissues, full of living cells called osteocytes. Each osteocyte is entombed in a hole called a lacuna, but is bathed in blood. In some parts of the bone, cells called osteoblasts are making new bone. Elsewhere osteoclasts are breaking it down.

Bones

- Bone can stand being squeezed twice as hard as granite and being stretched four times as hard as concrete.
- Weight for weight, bone is five times as strong as steel.
- Compact bone is the body's second hardest material after enamel.
- Bone is so light that it accounts for only 14 percent of our body weight.
- About three quarters of bone is compact bone; the rest is spongy bone.
- Even though calcium is important to the strength of bones, 99 percent of the body's calcium is in the teeth.
- When osteocytes die, their lacunae fill with salts and become important storehouses of minerals.
- The word collagen comes from the Ancient Greek "kolla," meaning glue and "gen," meaning forming. Collagen holds bone together.

MOVING SKELETON: JOINTS

One of the most remarkable things about the skeleton is that even though it is strong and rigid, it can bend and move in almost any direction when you want. This high degree of mobility depends on the skeleton's dozens of joints.

Joints are the places where bones meet. The joints in the skull are called fibrous joints. They are bound so tightly together with fibers that they are fixed firmly in place and cannot move.

All the other joints in the body are moveable. Some move easily in almost any direction. Others can move just a short way back and forth. Some are like hinges; other swivel like a ball in a socket.

Hinge joint

Saddle joint

Swivel joint

Ball and socket joint

Ball joint

HINGE JOINT
Hinge joints like those in the elbow swing only in two directions, like a door, but they are very strong.

SADDLE JOINT
Two saddle-shaped bone ends fit snugly together to allow great mobility but strength, as in the thumb.

SWIVEL JOINT
The swivel joint in between the skull and the spine allows you turn your head to the left and to the right.

BALL JOINT
Ball and socket joints like the shoulder and hip allow free movement in many directions.

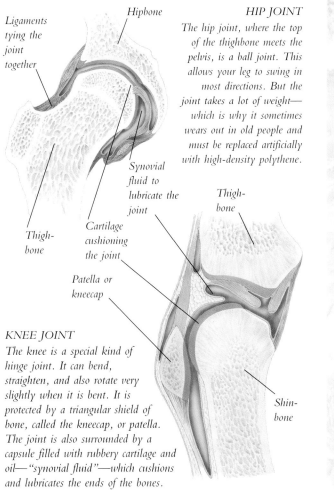

Ligaments tying the joint together

Hipbone

HIP JOINT

The hip joint, where the top of the thighbone meets the pelvis, is a ball joint. This allows your leg to swing in most directions. But the joint takes a lot of weight—which is why it sometimes wears out in old people and must be replaced artificially with high-density polythene.

Synovial fluid to lubricate the joint

Thigh-bone

Thigh-bone

Cartilage cushioning the joint

Patella or kneecap

Shin-bone

KNEE JOINT

The knee is a special kind of hinge joint. It can bend, straighten, and also rotate very slightly when it is bent. It is protected by a triangular shield of bone, called the kneecap, or patella. The joint is also surrounded by a capsule filled with rubbery cartilage and oil—"synovial fluid"—which cushions and lubricates the ends of the bones.

Joint protection

- Bones may take tremendous wear at the joints. So the body protects them with cartilage and synovial fluid.
- The ends of bones in most joints are covered in cartilage. Cartilage acts as a cushion between the bones because it is tough and rubbery. It is also very smooth, so it helps the joints to move together easily.
- Cartilaginous joints have only cartilage joining the bone. They are stiff but provide good support. The joints of the spine are cartilaginous.
- Synovial joints are joints that have synovial fluid as well as cartilage and move freely. Synovial joints are covered in a sleeve of tough collagen fiber. Inside the sleeve is the capsule of synovial fluid, a lubricant to match the best oils.
- When synovial joints like knees bend, the bone ends glide easily over the fluid.

JOINT ATTACHMENTS

Joints are the skeleton's weak points and so the body does everything it can to protect them. It provides rubbery cartilage to absorb shocks, oily synovial fluid to lubricate the joint, and special fibrous cords called ligaments to tie them together. Nearly all the major joints are supported by ligaments. Ligaments are attached to the bone on either side of the joint and help to anchor it in place. Ligaments are made of bundles of tough, rubbery collagen and elastin. They stretch a little to allow the joint to move, but they help to prevent it bending too far or twisting over. Ligaments, especially those in the ankle and knee, can be damaged by injury. A sprained ankle is a small tear or stretch in the ligaments supporting the ankle. The treatment is usually an ice pack, or sometimes a plaster cast.

SHOULDER JOINT

The shoulder gets a lot of wear and tear, and the bones on either side are protected by a coating of cartilage and separated by synovial fluid. The muscles are attached to the bones on opposite sides of the joint by stringy fibres called tendons, so that they pull the bones toward each other.

Cartilage

Synovial fluid

Shoulder bone

Tendon

Triceps muscle

Upper arm bone

Biceps muscle

KNEE LIGAMENTS AND TENDONS

The knee is made more sturdy when it bends by ligaments that prevent the knee from waggling to the side. The thigh muscle is anchored to the kneecap by tendons, and the kneecap is anchored in turn to the shin bone by ligaments.

Thigh muscle

Tendon

Kneecap

Meniscus (dish of cartilage)

Shinbone side ligament (tibia collateral ligament)

Shinbone (tibia)

Cruciate (crossing) ligaments

Calfbone side ligament (fibia collateral ligament)

Calfbone (fibia)

Joints

Kinds of cartilage

- There are three types of cartilage in the body: hyaline, fibrous, and elastic.
- Hyaline cartilage is the almost clear cartilage in joints and the ribs.
- Fibrous cartilage is the firm cartilage that forms the half-moon shaped cartilage around the knee that athletes sometimes damage.
- Elastic cartilage is more flexible and is used in airways, your nose, and ears.

Joint problems

- Joints are prone to damage and weakness, especially in old age.
- Arthritis is the inflammation of a joint.
- Rheumatoid arthritis is when the synovial fluid in small joints swells and thickens, making the joint stiff.
- Osteoarthritis is when the cartilage in a joint breaks down.

THE SKULL

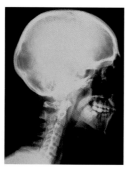

The skull is your head bone, the hard case that protects your brain, and provides a mount for your eyes, ears, nose, and mouth. It looks quite simple, but it is actually the most complicated part of the skeleton, made up of 22 different bones cemented together by rigid joints called sutures. The dome at the top encasing the brain is called the cranium, or cranial vault. The cranium is made of eight curved pieces of bone fused together along wriggly suture lines. The rest of the skull is the 14 bones that make up the face, including the lower jawbone, the skull's only moving bone.

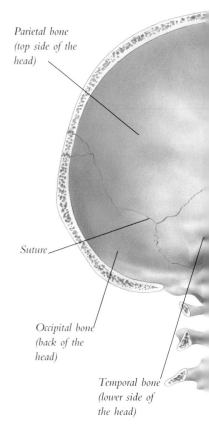

Parietal bone
(top side of the head)

Suture

Occipital bone
(back of the head)

Temporal bone
(lower side of the head)

SKULL X-RAY
No bones in your body give more of a clue to your personality than your skull. In the nineteenth century, people called phrenologists believed they could tell what kind of person you are from little bumps in your skull. This has no scientific basis. But archaeologists can reconstruct faces by computer analysis of ancient skulls.

76

HALF SKULL

This slice through the skull shows just some of its 22 bones. All the bones in your head are part of the skull—except for the three bones of each ear called ossicles. The bones of the face are shaped to create two bowl-shaped holes, called the orbits or eye-sockets.

Forehead
(frontal bone)

Sinus

Eye socket
(sphenoid bone)

Nose bone
(nasal bone)

Cheekbone
(zygomatic)

Upper jaw bone
(maxilla)

Lower jaw bone
(mandible)

The skull

Skull cavities

- The skull has four main cavities: the cranial cavity for the brain; the nasal cavity for the nose; and the two orbits for the eyes.

- A number of bones at the front of the face contain air spaces called sinuses. These lighten the skull and make the voice more resonant.

- Holes in the cranium allow blood vessels and nerves through, including the "optic nerves," the nerves from the eyes to the brain.

- The biggest hole in the cranium is in the base—the foramen magnum—where the brain stem goes out to join the spine.

Baby's skull

- A baby's skull bones are not fused together, and can slide and overlap to allow the head to squeeze through the birth canal. So a new born baby's skull has soft spots called fontanelles. The skull slowly fuses and hardens as the baby grows.

UPPER BODY BONES

The bones of the upper body focus on what is called the axial skeleton. This includes the skull, backbone, ribs, and breastbone. The arms and shoulders are suspended from this.

The spine is the remarkable bone structure that enables us humans alone to walk upright. It is actually a snakelike column of 33 drum-shaped bones called vertebrae, each connected by a joint at the back to allow a small amount of movement.

The ribs are the flat, curved bones that create a protective cage for the chest around the heart and lungs. There are seven pairs of "true ribs" attached to the breastbone, plus three pairs of "false" ribs, each attached to the rib above, and two pairs of "floating" ribs joined only to the spine.

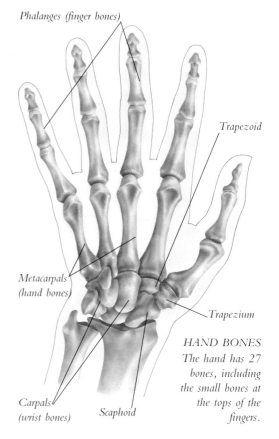

Phalanges (finger bones)

Trapezoid

Metacarpals (hand bones)

Trapezium

Carpals (wrist bones)

Scaphoid

HAND BONES
The hand has 27 bones, including the small bones at the tops of the fingers.

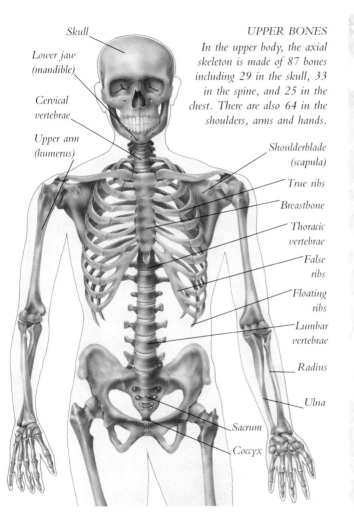

Skull

Lower jaw (mandible)

Cervical vertebrae

Upper arm (humerus)

Shoulderblade (scapula)

True ribs

Breastbone

Thoracic vertebrae

False ribs

Floating ribs

Lumbar vertebrae

Radius

Ulna

Sacrum

Coccyx

UPPER BONES

In the upper body, the axial skeleton is made of 87 bones including 29 in the skull, 33 in the spine, and 25 in the chest. There are also 64 in the shoulders, arms and hands.

The spine's sections

- The 33 vertebrae are separated by "disks" of jelly-filled cartilage, which cushion the bones when you run or jump.
- The top seven vertebrae are the neckbones that support the skull and are known as the cervical spine.
- The next twelve vertebrae run down the rear wall of the chest and there is a pair of ribs attached to each. This is the thoracic spine.
- The next five down make up the "lumbar" spine. This is the bit that carries weight when lifting.
- The five vertebrae below are fused together to make the "sacrum."
- The tail end of the spine is four vertebrae fused to make the "coccyx."
- In most people, the spine is S-shaped, with the cervical spine curving forward, the thoracic spine backward, the lumbar forward (especially in women), and the last bit backward.

LOWER BODY BONES

Your lower body bones are essentially your pelvis and the great long bones of your leg—the femur or thighbone, the tibia or shinbone, and the fibula or calfbone.

At the end of each leg, there are the feet. Remarkably, there as many bones in each foot as there are in your hand—26 altogether—even though your hand has to make more complicated shapes. But your foot has to support your body when you stand and act as a lever to propel you forward when you walk or run. So the foot's long bones are shorter, and the bones of the heel, which takes the weight, are very much bigger and sturdier than the equivalent bones in the hand.

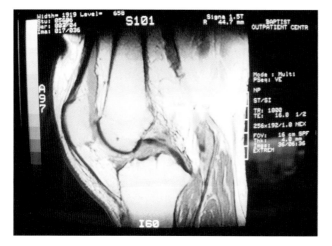

KNEE X-RAY
The knee joint takes an enormous amount of stress and strain. It is well protected by supporting ligaments, pads of cartilage, and lubricating synovial fluid (see p73). Even so, it is the most frequently injured joint in the body—especially among those who play sports such as football and skiing. This X-ray shows damage to the top of the tibia.

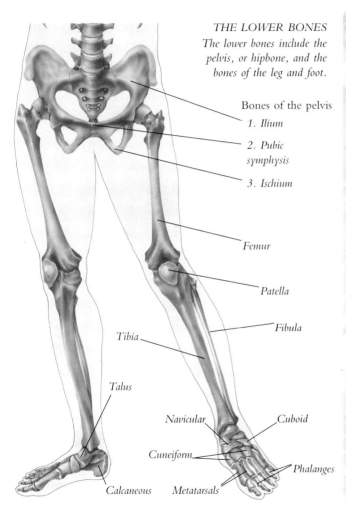

THE LOWER BONES

The lower bones include the pelvis, or hipbone, and the bones of the leg and foot.

Bones of the pelvis

1. Ilium

2. Pubic symphysis

3. Ischium

Femur

Patella

Fibula

Tibia

Talus

Navicular

Cuboid

Cuneiform

Phalanges

Calcaneous

Metatarsals

Leg and feet bones

Foot bones

- The foot and the toes are made from the same three sets of bones as the hand.

- The tarsals or heel bones are the equivalent of the carpals in the wrist.

- The metatarsals or instep bones are the equivalent of the metacarpals in the palm of the hand.

- The toes and fingers are both called phalanges.

- The tarsals are two bones. One of them—the heel-bone, or calcaneous—is the biggest bone of the foot.

Pelvis

- A girl's pelvis is much wider and shallower than a boy's. This is because the opening must be wide enough for a baby to get through as it is being born.

- The pelvis is made from three bones fused together: the ilium, the ischium, and the pubis.

- The ischium is the bone that takes all your weight when you sit down.

MUSCLES

Every move you make needs muscles. You need muscles to move your leg. You need muscles to twitch an eyebrow. You even need muscles just to sit still. Without muscles to hold you upright, you would slump like a sack of potatoes. And all their remarkable power depends on billions of tiny little sticky hooks working together.

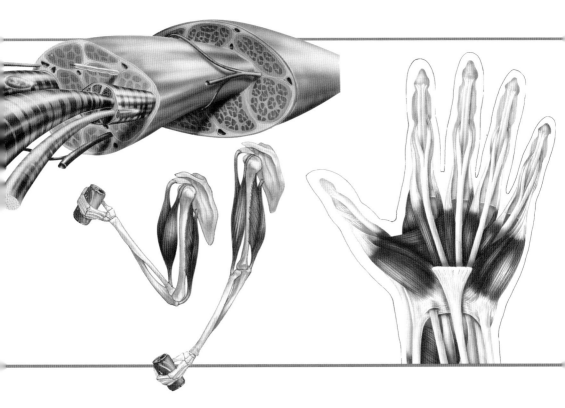

MUSCULATURE

Every move you make—running, dancing, smiling, and everything else—depends on muscles. You even need muscles to sit still. Without them would you would slump like a rag doll. Muscles are bundles of fibers that tense and relax to move different parts of the body, and there are two kinds: muscles that you can control, called voluntary muscles and muscles that you can't, called involuntary muscles. Most voluntary muscles are "skeletal" muscles that move parts of your body when you want. Involuntary muscles are like those of the heart that work automatically.

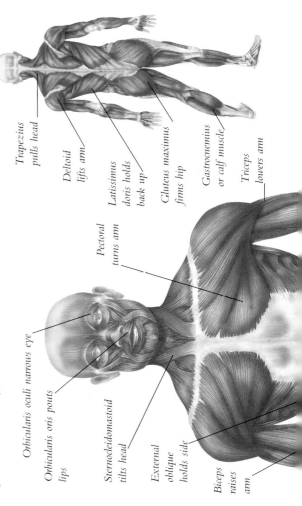

Trapezius pulls head

Deltoid lifts arm

Latissimus doris holds back up

Gluteus maximus firms hip

Gastrocnemius or calf muscle

Triceps lowers arm

Pectoral turns arm

Orbicularis oculi narrows eye

Orbicularis oris pouts lips

Sternocleidomastoid tilts head

External oblique holds side

Biceps raises arm

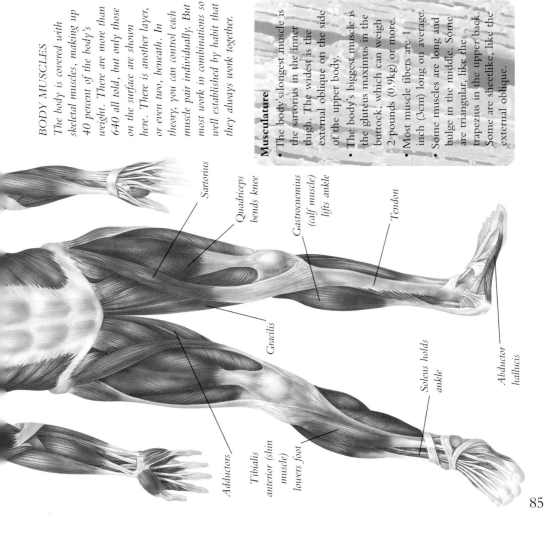

BODY MUSCLES

The body is covered with skeletal muscles, making up 40 percent of the body's weight. There are more than 640 all told, but only those on the surface are shown here. There is another layer, or even two, beneath. In theory, you can control each muscle pair individually. But most work in combinations so well established by habit that they always work together.

Musculature

- The body's longest muscle is the sartorius in the inner thigh. The widest is the external oblique on the side of the upper body.

- The body's biggest muscle is the gluteus maximus in the buttock, which can weigh 2 pounds (0.9kg) or more.

- Most muscle fibers are 1 inch (3cm) long on average. Some muscles are long and bulge in the middle. Some are triangular, like the trapezius in the upper back. Some are sheetlike, like the external oblique.

Sartorius

Quadriceps bends knee

Gastrocnemius (calf muscle) lifts ankle

Tendon

Gracilis

Soleus holds ankle

Abductor hallucis

Adductors

Tibialis anterior (shin muscle) lowers foot

85

KINDS OF MUSCLE

There are three kinds of muscle in the body: skeletal muscle, smooth muscle, and cardiac muscle. Skeletal muscles are the muscles covering your body, the "voluntary" muscles that you can control. They are sometimes called striated (stripy) muscles because they have dark bands around them. Smooth muscles and cardiac muscles are involuntary, which means that they work entirely automatically.

LEG MUSCLES
The muscles of your legs are voluntary muscles—muscles you can control. When you want to kick a ball, a message goes out from your brain via nerves. A fraction of a second later, the nerve signal reaches your leg muscles and makes them contract to move your leg and kick the ball.

SMOOTH MUSCLE

Smooth muscle drives movements inside the body. It propels food through the gut, for instance, which is why it is sometimes called visceral muscle. It also regulates the flow of blood through the arteries and veins. Smooth muscle forms tubes or sacs and gets its name because it is made in smooth flat sheets, rather than long bundles like skeletal muscle.

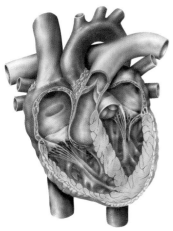

HEART MUSCLE
Cardiac muscle is the special muscle of which the heart is made. It is made of short interwoven branches of cells that spread the waves of muscle contraction that make the heart beat tirelessly.

CARDIAC MUSCLE

Cardiac muscle is the muscle of the heart that makes it beat. It is a combination of skeletal muscle and smooth muscle, but has its own built-in rhythm of about 100 beats a minute—but this can be varied by nerve signals from the brain. It is the one kind of muscle that must never tire, so it has to be fed huge amounts of oxygen and glucose in the blood.

GUT MUSCLE

Smooth muscle forms a tube around the gut. These muscles relax in front of lumps of food making the gut wider and contract behind, making it narrower. The effect is to push the food through the gut. This is called peristalsis.

Involuntary muscle

Smooth muscle

- Smooth muscle is made of loosely packed, tapering cells that contract slowly.
- Smooth muscle is often set in layers with fibers in different directions. In the gut, for instance, rings of muscle fibers are wrapped in a sheath of layers of lengthways fibers.
- Airways in the lungs are made of smooth muscle. In some people, these can go into spasm (contract sharply) due to an allergic reaction. This is one of the things that makes asthma sufferers short of breath. Salbutamol inhalers help by relaxing the muscles.

Cardiac muscle

- Cardiac muscle has specialized conducting cells for transmitting the nerve signals that spread waves of contraction through the heart (see p50).
- Heart muscle contraction begins in an area called the sinoatrial node on the side of the right atrium.

HOW MUSCLES WORK

The huge range of movements your body can make are all made simply by contracting and relaxing muscles. Each muscle moves the body just by pulling two points together.

This is why muscles must be anchored at both ends, attached to bones on either side of a joint, either directly or via tendons.

MUSCLE PAIRS

Muscles can shorten themselves, but they cannot make themselves longer. So each time a muscle contracts, it must usually be pulled back to its original length by another muscle shortening in the opposite direction. Muscles are therefore typically arranged in opposing pairs—a "flexor" to bend or "flex" a joint and an "extensor" to straighten it out again.

BENDING LEGS

When a skier does a fast turn on an icy slope, he uses his muscles to balance and to guide the skis. As he leans into the turn, he needs to transfer his weight onto his outside, downhill ski—the extra weight bends the ski in the middle so its edge carves the turn in the snow. To make the turn, his brain sends signals to contract the quadriceps at the front of one thigh and the hamstring at the back of the other. So his outside leg extends and pushes the weight on to this ski, while his inside leg flexes, lifting the weight off the other ski.

Biceps contracted to lift arm

Biceps relaxed

Triceps contracted to straighten arm

Triceps relaxed

ARM MUSCLES

In the upper arm, the muscles are attached to the top of the upper arm bone and the top of the lower arm bone. The biceps at the front is the flexor muscle and when it contracts and pulls on the lower arm, the arm bends at the elbow. The triceps at the back of the arm is the extensor muscle that pulls on the back of the lower arm to straighten the arm out again. Because the biceps have to do the lifting they are much bigger than the triceps.

Muscle movements

Moving and holding

- Not all muscle work involves movement; sometimes muscles contract to hold something still.
- When a muscle moves part of the body, it is called an isotonic contraction. Isotonic means "same force."
- When a muscle holds part of the body still, it is called an isometric contraction. Isometric means "same length," since the muscle does not actually get shorter.

Muscle contraction

- When muscles contract, they usually get fatter.
- Each muscle has a set of special nerve fibers that tell your brain how hard it is pulling when contracting. Another set in the tendon tells you how much it is being stretched when it relaxes and lets its partner pull the opposite way.
- Muscles don't always work in opposing pairs. They often work in complex combinations of 20 or more.

INSIDE A MUSCLE

Skeletal muscles owe their power to a special kind of cell. Muscle cells contain not just one nucleus like ordinary cells but ten or more, and they stretch from one end of the muscle to the other in a long fiber. Muscles are bundles of these fibers bundled together, like the fibers in string. Some muscles are made of just a few hundred fibers; others contain hundreds of thousands.

Just as the muscle is made from muscle fibers; so fibers are made up of even thinner strands called myofibrils. Under a powerful microscope, dark bands can be seen around the myofibrils. These line up so well they create stripes that give this kind of muscle its other name—stripy or "striated" muscle.

Actin filament

Myosin filament with hooks to pull the actin

MUSCLES POWER
Along myofibrils of muscles are dark and light stripes of two substances: actin and myosin. Actin and myosin filaments interlock. When nerve impulses tell the muscle to contract, little buds on each of millions of myosin filaments twist sharply, pulling together on the actin filaments and shortening the muscle.

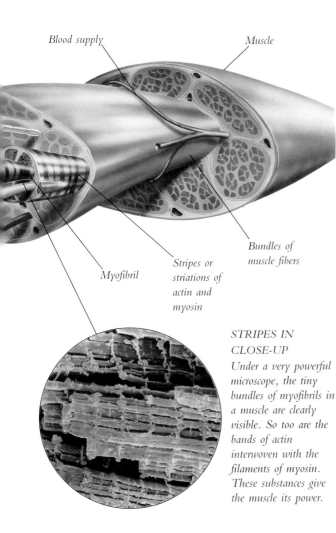

Blood supply

Muscle

Myofibril

Stripes or
striations of
actin and
myosin

Bundles of
muscle fibers

*STRIPES IN
CLOSE-UP*
*Under a very powerful
microscope, the tiny
bundles of myofibrils in
a muscle are clearly
visible. So too are the
bands of actin
interwoven with the
filaments of myosin.
These substances give
the muscle its power.*

Muscle power

- When contracting, many muscles can shorten by almost half their length.
- If all the muscles of your body pulled together, they could lift a big truck.
- 2,000,000 myosin filaments laid side by side would only be 1 inch (25mm) wide.
- The chemical "hooks" on myosin filaments are made of a stem called a cross-bridge and a head made of a chemical called adenosine triphosphate, or ATP.
- Every myosin filament has hundreds of cross-bridges, and the chemical process that makes muscle filaments pull together is called the cross-bridge cycle.
- When your brain sends a nerve signal to the muscle, a gush of calcium makes the ATP head on the myosin cross-bridges tilt toward the actin.
- When a muscle fires, billions of ATP heads tilt together in a fraction of a second, pulling the muscle shorter.

91

MUSCLE PERFORMANCE

Like a car engine, muscles need fuel, and the fuel they use is glucose. Muscles get energy when glucose joins with oxygen in a process called aerobic respiration. Sometimes when you start exercising, your muscles burn up glucose and oxygen so fast the blood can't deliver enough oxygen. So, for a while, the muscles burn glucose without oxygen. This is called anaerobic respiration (respiration without air).

SPRINTING HOME
During a long race, the muscles of an athlete can work aerobically most of the time. But the supply of oxygen in the blood may not be enough for the final sprint to the finishing line. When the dash begins, his muscles switch over to anaerobic respiration, drawing on his last reserves of internal energy for the final burst of power.

MUSCLE FIT
The body's capacity to supply oxygen to muscles increases by aerobic exercise—exercise that lasts long enough for muscles to work aerobically. Their lungs grow larger, and the heart beats more strongly and slowly.

92

SORE MUSCLES

If you're in shape, your heart soon starts pumping blood faster and blood vessels open up to boost the oxygen supply and restore aerobic respiration. If you are out of shape, your muscles go on working anaerobically much longer. This not only uses up glucose much more rapidly, tiring you out, but also leaves a build-up of lactic acid that makes overworked muscles feel sore. It is the body's efforts to draw in extra oxygen to burn off this acid that makes you pant when you stop running.

MUSCLE BUILDING
When muscles are exercised, they grow larger. At first the muscle fibers simply grow thicker. But continued regular training makes extra muscle fibers grow and the blood vessels supplying the muscle with oxygen grow more branches. So the muscles not only grow stronger but better able to keep going. This only happens, though, if the exercise pushes them to 80 percent of their capacity.

Muscle performance
- Energy consumption is measured in kilojoules or calories.
- Sitting down doing nothing, an average person burns up 7,000 kilojoules or 1,700 calories of energy.
- When you exercise hard, your energy consumption may double.

Chemicals in muscles
- Muscles get their energy by coiling up millions of molecules of the chemical ATP like a loaded spring. When it springs back, it pulls the muscle filaments together and shortens the muscle.
- In a resting muscle, energy is stored as a chemical called phosphoryl creatine. When the muscle starts to work, this chemical helps make coiled ATP.
- The store of phosphoryl creatine in each muscle is quite small. So if you go on exercising, the blood must supply glucose to replace it.

HAND MUSCLES

The human hand is capable of making an amazing range of movements—from fine and delicate movements like operating a computer keyboard or playing a piano, to strong and vigorous movements like wielding an ax or throwing a punch. Along with apes, humans are among the few animals able to grip things, because they can move their thumb and fingers independently.

The secret to the hand's combination of strength and delicacy lies in its combination of muscles and tendons. The small muscles that give it its fine movement are in the palm of the hand. But the big muscles that give it strength are in the forearm, and attached to the fingers and thumbs by tendons (cords).

Thumb-clenching muscle (adductor pollicis)

Short thumb-spreading muscle (abductor pollicis brevis)

Short thumb-bending muscle (flexor pollicis brevis)

Deep finger-bending tendon (flexor digitorum profundis)

Sheath for finger tendons

Upper finger-bending tendon (flexor digitorum superficialis)

Little finger-spreading muscle (abductor digiti minimi)

HAND MOVEMENT

When you grip firmly with your hand, the power comes from muscles in the forearm that shorten and bend the fingers by tugging on tendons. To cut friction to a minimum, these tendons are sheathed in a lubricating sleeve of synovial fluid. Delicate movements of your fingers, however, come from small muscles in your palm.

Left-handed and right-handed

- Nine out of ten people prefer to use their right hand for tasks such as writing, gripping a cup or tennis racket, or throwing a ball. These people are right-handed.

- One out of ten people is either left-handed or "ambidextrous"—that is, they can use both hands equally well.

- Handedness is nothing to do with the hand. It depends on which half of your brain is dominant. In right-handed people, it is the left half of the brain. In left-handed people it is the right half.

- In right-handed people, not only their hand is controlled by the left half of the brain, but their ability to speak too. In most left-handed people, the speech center is in the right half of the brain.

- Handedness is probably inherited from your parents.

UPPER BODY MUSCLES

The upper body includes the big, powerful muscles of the chest and arms as well as the many tiny muscles of the face that give us such an amazing range of expressions.

The upper back is dominated by a pair of large triangular muscles, called trapezius muscles. These are stretched over the back between the spine, neck, and shoulder blade. At the front are the muscles stretched across the upper chest called the pectorals.

The combination of trapezius and pectoral muscles are what give your shoulders such power. In fact, the shoulders are one of the weakest points of the skeleton, since the upper-arm bone is set in the shallowest of sockets, though you would hardly know it because it is supported and stabilized by six major muscles, including the deltoid.

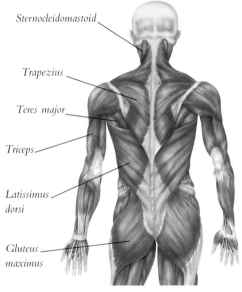

Sternocleidomastoid

Trapezius

Teres major

Triceps

Latissimus dorsi

Gluteus maximus

BACK MUSCLES
Because we human beings walk upright, our spines need tremendous muscular support. This is provided by two huge pairs of sheet-like muscles arranged in a cross over the back and wrapping right around to the front of the body. These muscles are called the latissimus dorsi and the gluteus maximus and they are the body's biggest muscles.

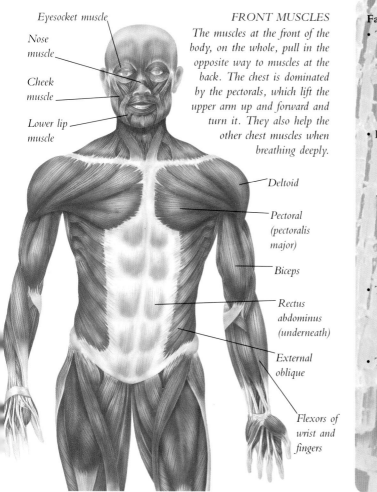

Eyesocket muscle

Nose
muscle

Cheek
muscle

Lower lip
muscle

Deltoid

Pectoral
(pectoralis
major)

Biceps

Rectus
abdominus
(underneath)

External
oblique

Flexors of
wrist and
fingers

FRONT MUSCLES

The muscles at the front of the body, on the whole, pull in the opposite way to muscles at the back. The chest is dominated by the pectorals, which lift the upper arm up and forward and turn it. They also help the other chest muscles when breathing deeply.

Face muscles

- The muscles of the face are small but complicated. Unlike other muscles, they are actually attached to the skin, so you can change your expression with just a slight muscle movement.

- Latin names for facial muscles include: chin (mentalis); lower lip (depressor labii inferioris); jaw (risorius); upper lip (levator labii superioris); lower cheek (zygomaticus major); upper cheek (zygomaticus minor); nose (nasalis); eye socket (orbicularis oculi).

- To smile, your upper lip muscle lifts your upper lip, while your cheek and jaw muscles pull the mouth up and out.

- To frown, the forehead muscle furrows your brow, your eyesocket muscles narrow your eyes, the nose muscle widens the nose, the lower lip muscle pulls down the lip, and the chin muscle puckers your chin.

LOWER BODY MUSCLES

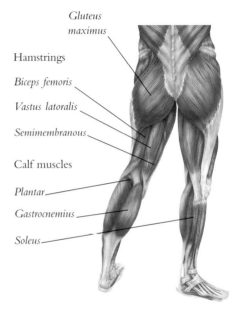

Gluteus maximus

Hamstrings

Biceps femoris

Vastus latoralis

Semimembranous

Calf muscles

Plantar

Gastrocnemius

Soleus

The lower body includes some of the body's biggest muscles— starting at the top with its very biggest pair, the gluteus maximus, the muscles of the buttock. The muscles of the lower body need to be big because they have to support all the weight of your body. They also need to be powerful to propel the body along when walking, running, and jumping.

The thigh is powered by big groups of muscles, not just two or three as the arm. There are biceps (double muscles) as in the arm. But the front thigh muscles—the muscles that straighten the knee—are "quadriceps." This means they have four parts. The calves, too, are powered by groups of muscles—the gastrocnemius, the soleus, and the plantar muscles. These are the muscles that raise and lower your heel.

LOWER-BODY MUSCLES: BACK
The muscles of the back of the lower body include the gluteus maximus, the powerful muscles of the buttock that move the thigh sideways and backward. They also include the hamstrings, a group of big muscles on the back of the thigh. These bend the knee and swing back your leg.

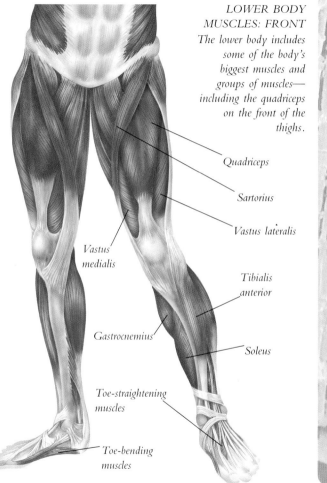

LOWER BODY MUSCLES: FRONT
The lower body includes some of the body's biggest muscles and groups of muscles— including the quadriceps on the front of the thighs.

Quadriceps

Sartorius

Vastus lateralis

Vastus medialis

Tibialis anterior

Gastrocnemius

Soleus

Toe-straightening muscles

Toe-bending muscles

Leg muscles
Tendons
- The big muscles of the legs are attached to their bones by extra strong tendons.
- The calf muscles pull up the back of the heel through a very strong tendon, called the Achilles tendon.
- The Achilles tendon is named after Achilles, a legendary hero of Ancient Greece, whose only weakness was his heel.
- Runners often damage their Achilles tendon through poor running technique or wearing the wrong shoes.
- Tendons are attached to bones by fibers embedded deep within the bone. These are called Sharpey's fibers or perforating fibers.
- After a tendon is injured, it may become inflamed. This is called tendinitis.
- Typists often suffer from tenosynovitis in their wrists. This is inflammation of the tendon's sheath, and it is caused by excessive small movements of the fingers.

FOOD AND EATING

Like any machine, your body needs fuel to keep it going—and that is the main reason you need to eat. Food is the body's main source of energy. It is also a source of many other vital materials for the body. But food provides neither energy nor materials in the form the body needs, so inside you, after every meal, an extraordinary chemical processing factory gets to work…

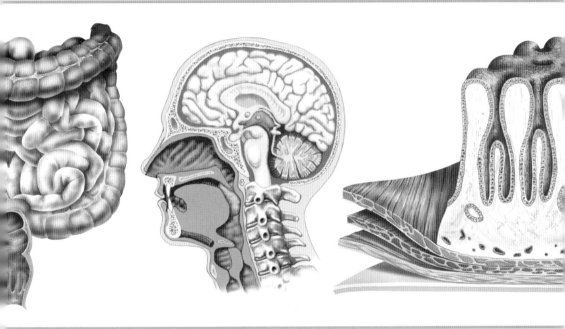

DIGESTIVE SYSTEM

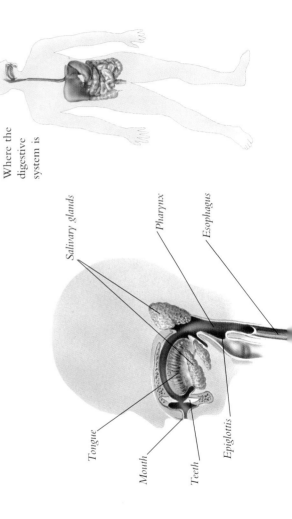

The body needs food in the form of small, simple molecules that can be delivered to cells in the blood. Yet the food you eat comes in big lumps and large complex molecules. So the body has its own food refinery for breaking food down into the right kind of molecules. This food breakdown refinery is known as the digestive system. It is essentially a long tube through the body, called the alimentary canal, through which the food slowly passes and is gradually digested—that is, broken down into small molecules and absorbed into the bloodstream.

Where the digestive system is

Salivary glands

Pharynx

Esophagus

Tongue

Mouth

Teeth

Epiglottis

THE DIGESTIVE SYSTEM

The digestive system is essentially the alimentary canal, plus the liver and pancreas. The alimentary canal begins in the mouth, runs down through the esophagus, or throat, then the stomach, the small intestine and the large intestine and finally ends at the anus.

The digestive system

- The alimentary canal is more commonly known as the gut.
- Your gut is folded over many times. If you could lay it out straight, it would be nearly six times as long as you are tall!
- When you are sick, your diaphragm and the muscles of your abdomen contract to squeeze partly digested food from your stomach out through your mouth. Vomit tastes bitter because of the acidic stomach juices it contains.

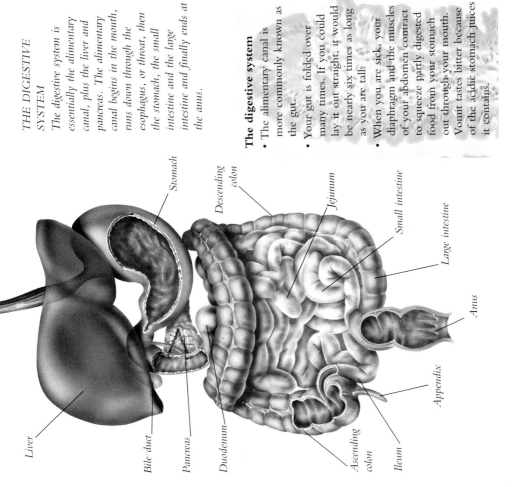

Stomach

Descending colon

Jejunum

Small intestine

Large intestine

Anus

Appendix

Ileum

Ascending colon

Duodenum

Pancreas

Bile duct

Liver

EATING

The process of food breakdown begins the moment food enters your mouth. Your jaw muscles are very powerful—the most powerful for their size in your body—and they grind the teeth together with huge force. Aided by the dissolving power of enzymes in saliva, they soon chew the food to a soft pulp.

When the food is broken up enough to be swallowed, the tongue pushes it back toward the throat. The mouth of the roof or palate rises to block off the passage to the nose, and the larynx is shut off by the epiglottis. The lump of food, or "bolus," slips into the gullet or "esophagus," the tube that leads down to the stomach.

SWALLOWING
This series of illustrations shows how the roof of the mouth rises and the airways are blocked off so that a bolus of chewed food can be swallowed without choking you.

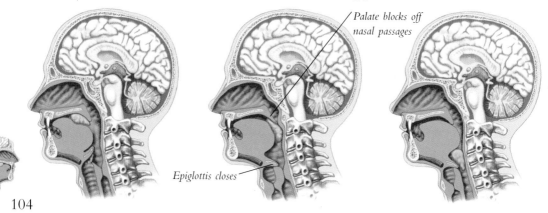

Palate blocks off nasal passages

Epiglottis closes

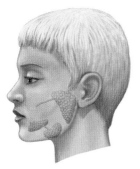

SALIVARY GLANDS
When your mouth waters, salivary glands are oozing saliva into your mouth. Saliva not only helps lubricate food and makes it easier to swallow, it also contains amylase, a chemical enzyme that helps dissolve food.

Chewing and swallowing

- Although you can choose when to swallow, once started the whole process is automatic.
- Once swallowed, powerful muscles in the esophagus squeeze food down to the stomach in under two seconds.

Salivary glands

- There are three main pairs of salivary glands in the mouth: the parotid, the submandibular, and the sublingual.
- There are also many small salivary glands all around the mouth.
- The parotid glands, just in front of the ears, are the largest. Their ducts open into the mouth from the cheek near the molars.
- The submandibular glands are at the back of the lower jaw, under the mouth, and ducts enter the mouth under the middle of the tongue.
- The sublingual glands are under the tongue.

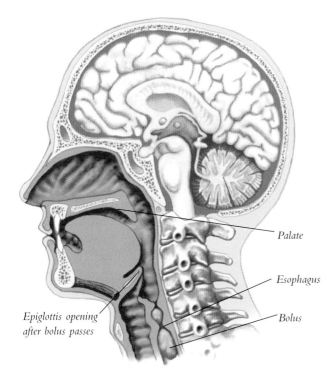

Palate

Esophagus

Bolus

Epiglottis opening after bolus passes

105

THE STOMACH

The stomach is where the body's breakdown of food really begins. The stomach has strong, muscular walls. As soon as food enters, waves of squeezing and relaxation begin to sweep across the stomach, churning and pounding the food. At the same time, glands in the stomach secrete "gastric" juices, including acids and enzymes, that start to dissolve the food chemically. To protect the stomach from its own juices, the stomach walls are lined with a thick layer of mucus. Sometimes, this lining can break down, and the acids attack the stomach wall, causing a painful stomach ulcer.

THE STOMACH
The stomach is a large curved bag situated in the middle of the chest, a little way above your navel. It is just behind the liver. You can see that the exit is almost as high as the entrance so that food does not weigh down on the pyloric sphincter. The illustration far right shows the stomach lining greatly magnified.

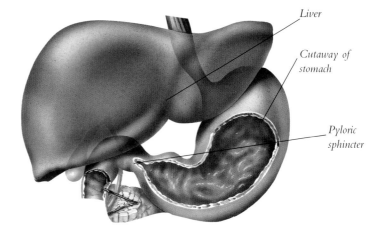

Liver

Cutaway of stomach

Pyloric sphincter

WAITING ROOM

As well as breaking food down, the stomach also acts as a storehouse for the partially digested food, and lets it through only gradually to the next stage of the digestive system, the small intestine. The stomach is separated from the small intestine by a ring of muscle called the pyloric sphincter. This acts like a rubber band round the neck of a bag, tightening or relaxing, to control the amount of food that passes through.

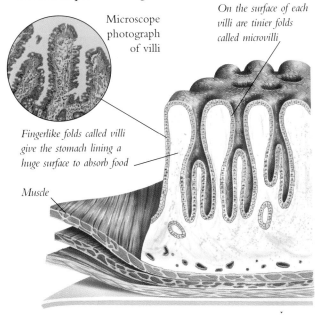

Microscope photograph of villi

On the surface of each villi are tinier folds called microvilli

Fingerlike folds called villi give the stomach lining a huge surface to absorb food

Muscle

The stomach

- When empty, the stomach is quite small, holding barely a pint (450ml). But after a heavy meal, it may stretch to hold more than 8 times as much.

- When the stomach is full, waves of muscle contractions sweep across it every 20 seconds or so.

- One of the acids secreted by the stomach is concentrated hydrochloric acid—which is why the stomach must be lined with mucus. The acid kills most bacteria and other dangerous microorganisms and helps dissolve food.

- The richer the food, the longer it takes to empty from the stomach.

- If your stomach didn't store food, you would have to eat every 20 minutes or so.

- Stomachache or indigestion is pain anywhere in the intestine, often caused by eating too much, too quickly, or by eating food that it is very rich or spicy.

THE GUT

By the time partially digested food leaves the stomach and passes into the small intestine, it is in a semiliquid mass called chyme. Over the huge length of the small intestine—more than 23 feet (7m)—the chyme is broken down further. Eventually it is broken into small molecules that can seep through the walls of the intestine and into the bloodstream. The blood then carries it to the liver for processing.

Most food is digested in a section of the intestine called the duodenum. It is absorbed into the blood in the "ileum." Food that cannot be digested or absorbed passes on into the large intestine and is expelled from the body through the anus.

PERISTALSIS
Food is moved through the gut by the muscles of the gut wall. Whenever a chyme of food enters the gut, it stimulates waves of muscle contraction and relaxation. Rings of muscle just behind the chyme of food contract sharply, while muscles in front relax. So the chyme is eased gradually forward, as the waves pulse down the tube. This is called peristalsis. Peristalsis not only helps move food, but mixes and batters it ready for chemical assault.

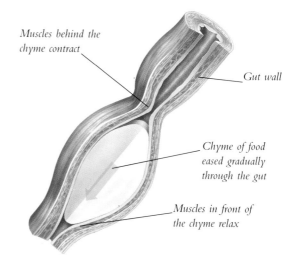

Muscles behind the chyme contract

Gut wall

Chyme of food eased gradually through the gut

Muscles in front of the chyme relax

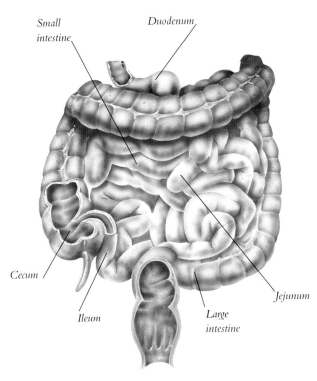

Small intestine

Duodenum

Cecum

Ileum

Large intestine

Jejunum

THE SMALL AND LARGE INTESTINE

The intestine, or gut, is divided into two sections: the narrower small intestine where food is digested and absorbed, and the wider large intestine where undigested food is dried out and prepared for evacuation. To perform its task thoroughly, the gut has to be incredibly long, so it is folded over and over again inside your abdomen.

Digestion timetable

- Food takes on average 24 hours to pass through the alimentary canal (see p102) and out the far end.

- *6 PM* As you swallow your dinner, it travels down into your stomach, where it is gradually broken down into semiliquid chyme.

- *10 PM* Partially digested food passes through the pyloric sphincter as chyme into the duodenum of the small intestine. Here it is fully digested.

- *1 AM* Digested food passes through the jejunum of the small intestine to the ileum where useful molecules are absorbed into the blood.

- *3 AM* Any waste passes into the colon of the large intestine, where water is absorbed from the chyme.

- *11AM - 5PM* The waste passes into rectum as feces and is eventually excreted through the anus, after anything from one to three days.

DIGESTION

The digestion of food is not simply a mechanical process, like chewing, but involves all kinds of chemicals. Acids such as bile are important, but so too are special chemicals called enzymes. Digestive enzymes do not actually break up food themselves; they simply get processes going. But the digestive enzymes are like biological scissors, snipping away at the giant molecules of food like kitchen scissors on a string of sausages. The enzyme amylase, for instance, chops up molecules of starch in bread, potatoes, and other vegetables into simpler sugars.

DIGESTING CARBOHYDRATES

This illustration shows how carbohydrates are broken down by various enzymes first into simple sugars, then glucose so that they can be absorbed into the blood. The liver stores some glucose as glycogen, but releases the rest for immediate use by the body. The hormones insulin and glucagon, made by the pancreas, tell the liver when to release glucose and when to hold onto it.

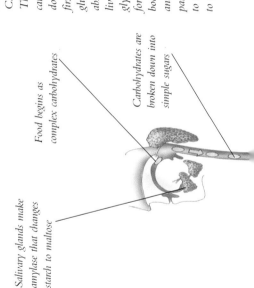

Food begins as complex carbohydrates

Carbohydrates are broken down into simple sugars

Salivary glands make amylase that changes starch to maltose

Insulin tells the liver to store glucose as glycogen

Glucagon tells the liver to change glycogen into glucose

When blood sugar is low, the pancreas sends the hormone glucagon to the liver

When blood sugar is high, the pancreas sends the hormone insulin to the liver

Glucose goes to the liver

The liver stores some glucose as glycogen

Another version of amylase is secreted by the pancreas into the duodenum

In the duodenum, amylase cuts down carbohydrates into simple sugars—maltose, lactose, and sucrose

The enzymes maltase, lactase, and sucrase snip maltose, sucrose, and lactose molecules into bits

Simple sugars are broken down into glucose

Glucose is absorbed into the bloodstream

Digesting proteins and fats

- The stomach wall oozes out an enzyme called pepsinogen.

- Pepsinogen has little effect, but when it joins with stomach acid, it becomes pepsin that breaks protein in food into chains of amino acids.

- Amino acid chains are snipped smaller in the small intestine by the enzymes trypsin and peptidases.

- Fats are broken in the small intestine by bile from the liver and the enzyme lipase.

A HEALTHY DIET

Your body is remarkably good at taking care of itself. But to stay healthy, you need to eat the right food. Most of the food you eat is fuel burned by the body for energy, and it gets this mainly from substances the body extracts from food called carbohydrates and fats. But your body also needs small quantities of foods such as proteins, which are needed to repair cells and build new ones. It also needs minute traces of chemicals that it cannot make for itself—vitamins and minerals. A healthy diet is one that includes all of these foods in just the right proportion and quantity.

MEAT

Protein in fish and meat has all the amino acids the body needs. Fruit and vegetables have only some of them, which is why vegetarians must choose the right mixture. Meat is also rich in saturated fats.

Vegetables are important sources of vitamins

Unlike white bread, whole-wheat bread is good for fiber

Cheese is 25% protein, but also rich in saturated fats

Foods such as bread, rice, potatoes, and sweet things are rich in carbohydrates

ESSENTIAL FIBER

To keep the muscles of your bowel exercised, you need to eat "roughage"—cellulose plant fibers that your body can't digest. Beans, fruit, and whole-wheat bread are all good sources of roughage.

Bananas are a good source of instant energy

Fish and eggs are good sources of protein and vitamins

FOOD BASKET

Besides vitamins and minerals, a good diet includes just the right quantities of each type of food—carbohydrates, fats, and proteins. Too much of any can be as bad as too little. Too much fat, for instance, not only makes you fat—too much "saturated" fat (mainly animal fat such as butter) can also increase the risk of heart disease.

Body foods

Carbohydrates

• Carbohydrates are foods made of kinds of sugar like glucose and also starch, which is turned into sugar in your body.

• If you eat more carbohydrates than you need they are changed to glycogen and stored in muscles and the liver.

Fats

• Fats are greasy foods that won't dissolve in water. Some are solid, like meat fat and cheese. Others are liquid, like corn oil.

• The body usually stores fats for future energy use rather than burning it at once.

Proteins

• The body needs 20 "amino acids" to build into protein to make and repair cells. The body can make 12 of these; the other eight it must get from protein in food. These are the "essential amino acids."

• The faster you are growing, the more protein you need.

VITAMINS

Besides carbohydrates, proteins, and fats, the body also needs from food tiny traces of chemicals it cannot make itself: vitamins and minerals. Vitamins are needed to help drive various processes inside cells. When they were first discovered, they were given letter names, such as Vitamin B. But later discoveries were given chemical names. Each is found in certain foods and has its own range of tasks in the body. Essential minerals include salt for maintaining water levels, calcium for bones, iron for red blood cells, tiny traces of iodine, and various other minerals.

VITAMIN CRYSTAL
This microscope photograph shows a crystal of Vitamin C magnified many times. Some vitamins can be dissolved in fat, such as A, D, E, and K. The body absorbs them from fats and they are then stored for many months. Others dissolve in water, such as C and the Bs, and are needed daily.

Vitamin	Tasks in the body	Food sources
Vitamin A or retinol	• Regulates growth of skin cells, helps regeneration of cells in eyes and lungs.	• Liver, fish, dairy products, eggs, carrots, and green vegetables, especially spinach.
Vitamin B or thiamine	• Vital for digestion, growth, and muscle tone and for getting energy from glucose. Reduces pain.	• Most carbohydrate-rich foods such as bread, flour, and potatoes, plus meat, milk, peas, beans, and beer.
Vitamin B_2 or riboflavin	• Helps change food to energy.	• Milk, liver, eggs, vegetables.
Vitamin B_6 or pyroxidine	• Helps body deal with amino acids and form haemoglobin in red blood cells.	• Most foods, especially liver, cereals, bread, dairy products, and eggs.
Vitamin B_{12} or cyancobalamin	• Good for nervous system; and for bone marrow.	• Dairy products and meat, especially liver.
Biotin	• Helps enzyme systems, cell growth, and making fats.	• Egg yolk, liver, kidney; made by bacteria in the gut.
Folic acid	• Helps bone marrow and nerve defects in fetuses.	• Leafy vegetables, liver, pulses, bread, oranges, bananas.
Nicotinic acid or niacin	• Helps nerves and good for skin and appetite.	• Lean meat, cereals and bread, eggs, dairy products.
Panthothenic acid	• Needed for growth, making antibodies, and digesting fat.	• Panthothenic acid is found in nearly all food.
Vitamin C or ascorbic acid	• Vital for teeth, gums, and bones; helps the body fight infection and deal with toxic substances.	• Fresh fruit, especially oranges, lemons, and cranberries, plus green vegetables and potatoes.
Vitamin D or calcifol	• Important for maintaining calcium in blood and bones.	• Cod liver oil, egg yolk, salmon, butter, plus sunlight.
Vitamin E or tocopherol	• Antioxidant, keeps walls of red blood cells healthy.	• Vegetable oils, eggs, nuts, fruits, and green vegetables.
Vitamin K or phylloquinone	• Vital for blood clotting.	• Green vegetables, oats.

THE CHEMICAL FACTORY

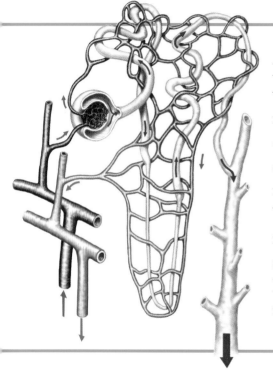

Between your navel and your backbone, every minute of the day, one of the universe's most amazing little chemical factories is at work. Your liver is performing miracles of chemical synthesis, while your kidneys purify blood on the move.

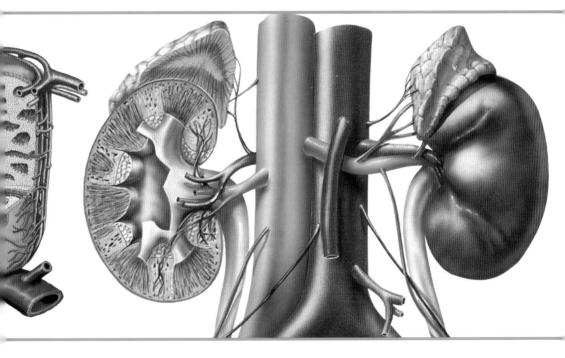

HOW THE LIVER WORKS

The liver is the body's biggest organ—and one of its cleverest, performing miracles of chemical processing on many different substances. It helps purify the blood, for instance, sweeping out all the tired old red cells and poisons such as alcohol. It does not actually make blood cells, but it makes vital proteins for blood plasma and is also a source of bile, the fluid that helps dissolve fat in food.

The liver's most important task, though, is to receive chemicals digested from food, and repackage them for use all around the body when needed. It takes carbohydrates, for instance, and turns them into glucose, the number one fuel of body cells. Guided by two chemical messengers, glucagon and insulin, the liver helps ensure our blood always has the right level of glucose—without which we would fall into a coma and quickly die.

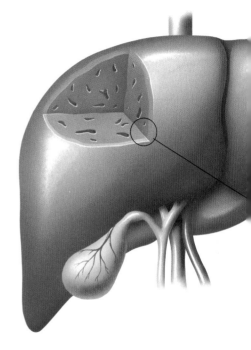

LIVER
The liver is a large organ sitting just above the stomach. Despite the huge range of tasks it performs, it is actually quite simply made. Its jellylike interior is made up from thousands and thousands of hexagonal lobes, or "lobules."

118

LIVER LOBE

The key to the liver's operations are the box-shaped "lobules," with sections like the segments of oranges. Blood from the liver's twin supplies flow into each segment from the outside edge, through a channel called a sinusoid, and out through a vein in the center. There are special liver cells or "hepatocytes" lining the sinusoid, and as the blood flows through it, these cells extract the right chemicals, process them, and return them to the blood—except for bile, which is sent out the back.

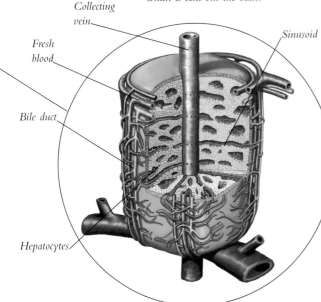

Collecting vein

Fresh blood

Bile duct

Hepatocytes

Sinusoid

Liver facts
- The word "hepatic" means belonging to the liver.
- Uniquely, the liver has two blood supplies. One is an artery, the hepatic artery; the other is a vein, the portal vein. Both enter the liver through its "portal," or gate.
- The third pipe passing through the liver's portal is the common "bile" duct, which carries bile, the acid the stomach uses to help break fatty food down.
- The liver acts as a despatch center for the chemicals extracted from food by the digestive system.
- The liver is the body's prime energy store, holding glucose in the form of the chemical glycogen.
- The liver packs off excess food energy for long term storage as fat.
- The liver breaks down proteins and stores vitamins and minerals.
- The liver clears the blood of old blood cells and makes new plasma and proteins.

119

INSIDE THE KIDNEYS

Water is crucial to the working of the body, and the kidney is the key to water control. It holds water back when needed, or lets it run out as urine when there is too much. The water is mixed in with the blood, so the kidney has to draw off water without losing any of the blood's vital ingredients. At the same time, it must clear the blood of poisonous waste. The kidneys are basically a highly efficient filtration unit, cleaning the blood as it washes through—catching larger materials and letting smaller blood ingredients pass through to the next stage. The kidney then reabsorbs the wanted ingredients and water, and allows toxic waste and unwanted water to flow away in the urine.

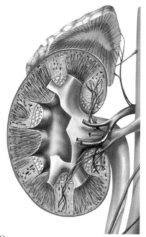

KIDNEY

The kidneys are a pair of bean-shaped organs in the small of the back, along the body's main arteries and veins. Blood entering the kidney through an artery is distributed through a million or so filtration units called nephrons.

NEPHRON

A nephron is an intricate network of little tubes called convoluted tubules, wrapped around by an even more intricate network of blood capillaries. It is in these capillaries that wanted ingredients are reabsorbed into the blood.

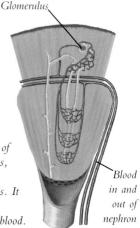

Glomerulus

Blood in and out of nephron

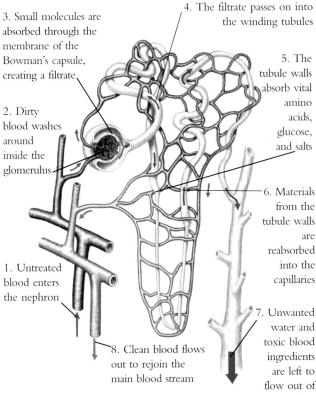

3. Small molecules are absorbed through the membrane of the Bowman's capsule, creating a filtrate

4. The filtrate passes on into the winding tubules

2. Dirty blood washes around inside the glomerulus

5. The tubule walls absorb vital amino acids, glucose, and salts

1. Untreated blood enters the nephron

6. Materials from the tubule walls are reabsorbed into the capillaries

8. Clean blood flows out to rejoin the main blood stream

7. Unwanted water and toxic blood ingredients are left to flow out of the tubule as urine

NEPHRON IN CLOSE UP

Blood is fed into each nephron through a little bundle of capillaries called the glomerulus. This is held in a cup called the Bowman's capsule. Between the two there is a thin membrane, and this membrane is the kidney's filter.

Kidneys

- Blood flows through the kidneys at 3 pints (1.3 l) a minute.
- All the body's blood flows through the kidneys in just 10 minutes, so the blood is filtered by the kidneys hundreds of times a day.
- The kidneys absorb 370 pints (170 l) of filtrate from 4,400 pints (2,000 l) of blood everyday.
- The kidneys only let go 3 pints (1.2 l) of urine from every 4,400 pints (2,000 l) of blood, reclaiming all the rest.
- Only small molecules get through the membrane of the Bowman's capsule— salt, water minerals, glucose, urea, and creatine.
- Urea is the waste from the breakdown of proteins in the liver; creatine is the waste from muscle action.
- The tubule walls retrieve all of the vital amino acids (see p113) and glucose from the filtrate for returning to the blood, and 70 percent of the salt.

121

THE WATER SYSTEM

Your body needs one more vital input as well as oxygen and food—water. You can live for a month without food but only a few days without water. If the body's water content goes up or down by 5 percent, the results are disastrous. The urinary system plays a key role in keeping the amount of water in the body steady.

You gain water by drinking, eating, and as a by-product of cell activity. You lose it by sweating, breathing, and through urinating. The amount you gain and lose through cell activity, sweating, and breathing stays much the same. So your body controls water by balancing the amount you drink against the water you lose in urine.

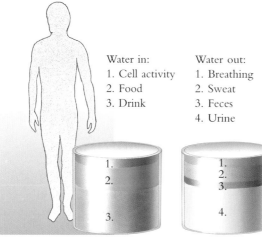

Water in:
1. Cell activity
2. Food
3. Drink

Water out:
1. Breathing
2. Sweat
3. Feces
4. Urine

WATER IN THE BODY
Typically, you will take in 5 pints (2.2 l) of water a day, 3 pints (1.4 l) in drinks and 2 pints (0.8 l) in food. Body cells add an extra 11 ounces (0.3 l), bringing the total to 5 pints, 11 ounces (2.5 l). So, to keep you from being swamped, your body must also lose this amount. Typically, 11 ounces (0.3 l) goes out as vapor on your breath, 18 ounces (0.5 l) in sweat, 7 ounces (0.2 l) in your feces, and 2 pints, 12 ounces (1.5 l) in

1. Water is drawn off from the blood in the kidneys

2. It trickles down the kidney's tubules and into the ureter

3. Urine collects in the bladder

4. The pressure of urine on the ring of muscle at the exit of the bladder builds up

5. You become aware that your bladder is full and find a place and time to urinate

KIDNEY WATER

The urinary system drains unwanted water drawn off from the blood by the kidneys. It pipes the water or "urine" down to the bladder, where it builds up until the pressure of water there makes you want to urinate.

Water and salts

How the kidneys control water

• The amount of water the kidneys let out as urine depends on the amount of salt dissolved in the blood.

• If you drink a lot, the water in the blood becomes very diluted, and the kidneys let lots of water out, giving pale and watery urine.

• If you drink only a little or sweat a lot, the kidneys hold on to more water, and the urine is stronger.

Why water balance matters

• There is slightly more potassium salt dissolved in the water inside body cells than in the water outside.

• There is slightly more sodium chloride (common salt) dissolved in the water outside cells.

• Water seeps through cell walls from areas of low salt to areas of high salt in a process called osmosis.

• If the body water gets too dilute, it seeps into cells and swells them up. If it gets too salty, cell water leaks out.

123

EXCRETION

Your digestive system is remarkably good at breaking down the food it receives and taking what it wants. But there are some things that it can do nothing with. It is the task of the final part of the digestive system—the large intestine—to deal with this unwanted food waste. The winding 5 feet (1.5m) of your colon changes this waste into a suitable form for the body to excrete as feces through your rectum and anus.

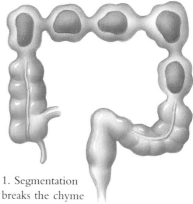

1. Segmentation breaks the chyme into short segments

FULL OF BEANS

The efficient functioning of the colon and bowel depends on plenty of roughage—fibrous material made from plant cellulose, which the body finds indigestible. This fiber holds the feces together and keeps the bowel exercised. So it is good to eat food with plenty of roughage in it, including beans and whole wheat bread.

DRYING OUT

The main task of the colon is to convert the liquid chyme passed on by the small intestine into solid faeces. It absorbs a lot of water and some of the salt from the chyme to dry it out. The colon is helped in its task by billions of bacteria which live inside the intestine. They feed on undigested fiber and make vitamins K and B, as well as gases such as hydrogen, methane, and hydrogen sulfide.

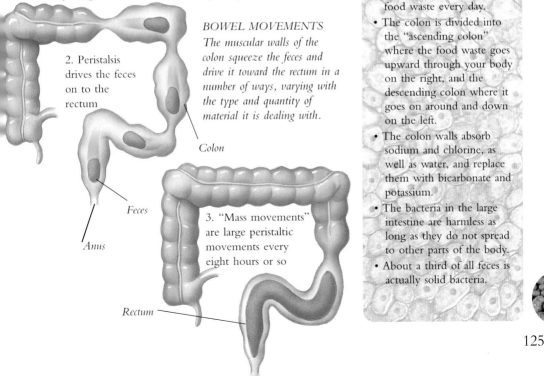

2. Peristalsis drives the feces on to the rectum

BOWEL MOVEMENTS
The muscular walls of the colon squeeze the feces and drive it toward the rectum in a number of ways, varying with the type and quantity of material it is dealing with.

Colon

Feces

Anus

3. "Mass movements" are large peristaltic movements every eight hours or so

Rectum

The colon and below

- Chyme enters the large intestine through a little trapdoor called the ileocecal valve, which opens to let a bit through at a time.
- The first part of the large intestine, before the colon, is called the cecum.
- The colon absorbs 3 pints (1.3 l) of water from the food waste every day.
- The colon is divided into the "ascending colon" where the food waste goes upward through your body on the right, and the descending colon where it goes on around and down on the left.
- The colon walls absorb sodium and chlorine, as well as water, and replace them with bicarbonate and potassium.
- The bacteria in the large intestine are harmless as long as they do not spread to other parts of the body.
- About a third of all feces is actually solid bacteria.

THE NERVOUS SYSTEM

Nerves are your body's hot lines, carrying instant messages from the brain to every organ and muscle in the body—and sending back a constant stream of information to your brain about the world both inside and outside your body.

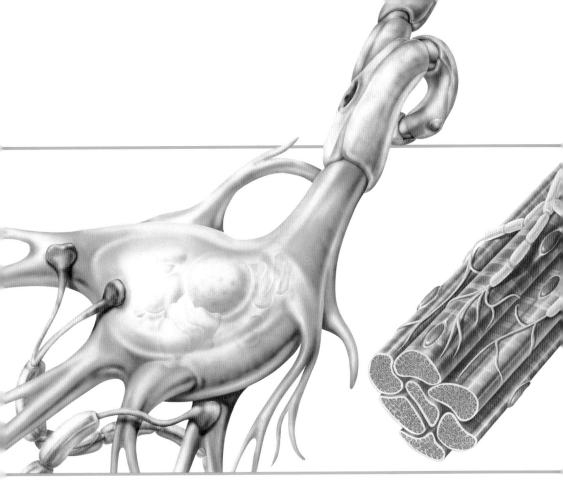

NERVOUS SYSTEM

Like a busy telephone network, your nervous system is continually buzzing with activity, whizzing messages to and fro all over the body. All the time, millions of nerve impulses reach the brain from the body's sense receptors, and just as many leave the brain with instructions for muscles to move or organs to work.

The heart of your nervous system is your brain and the bundles of nerves running down your spine, known as the spinal cord. Together, the brain and spinal cord make the central nervous system or CNS. Nerves spread out from the CNS to all parts of the body. These nerves are the Peripheral Nervous System or PNS.

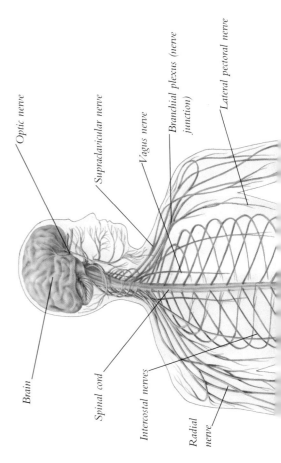

Optic nerve

Supraclavicular nerve

Vagus nerve

Branchial plexus (nerve junction)

Lateral pectoral nerve

Brain

Spinal cord

Intercostal nerves

Radial nerve

THE NERVOUS SYSTEM
Nerves link your brain to every part of the body. Like branches on a tree, the nerves of the PNS spread out in all directions from the CNS (the brain and spinal cord) in all directions. The main branches of the PNS are the 12 "cranial" nerves in the head and the 31 pairs of "spinal" nerves that branch off the spinal cord. All the other nerves branch off these.

The ANS

- Besides the CNS and PNS, the body has a third nervous system, the autonomic nervous system or ANS, which controls automatic functions such as heartbeat and digestion.

- Many cells in the ANS are in small groups called ganglia, which monitor and control things like glands by themselves.

- The ANS has two parts: the parasympathetic, which deals with everyday tasks; and the sympathetic, which prepares the body for action.

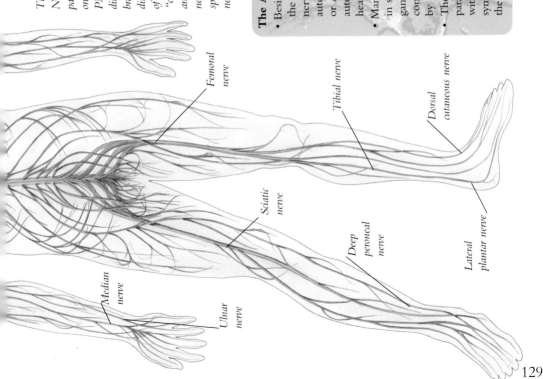

Femoral nerve

Tibial nerve

Dorsal cutaneous nerve

Sciatic nerve

Deep peroneal nerve

Lateral plantar nerve

Median nerve

Ulnar nerve

129

How Nerves Work

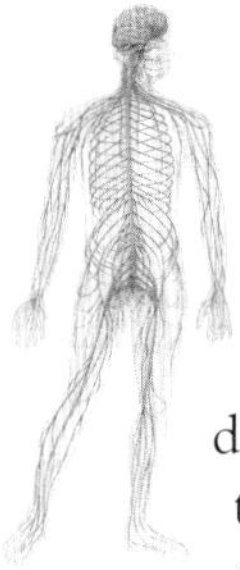

Your nervous system is made up from long strings of special nerve cells or neurons—all strung together like beads on a string. Most cells are like small parcels, with a nucleus at the center, like an egg yolk. But neurons are a very odd spidery shape. Nerve signals enter the neuron through tiny branching threads called dendrites. They then rush along the cell's very long, thin winding tail or axon. Axons pack together like the threads in string to make nerve fibers. At the far end, called the interneuron, the axon branches into hundreds more feathery dendrites that pass the message on to the dendrites of other cells.

In a way, a nerve is rather like an electric wire. But nerve signals are actually sent by a remarkable mix of electricity

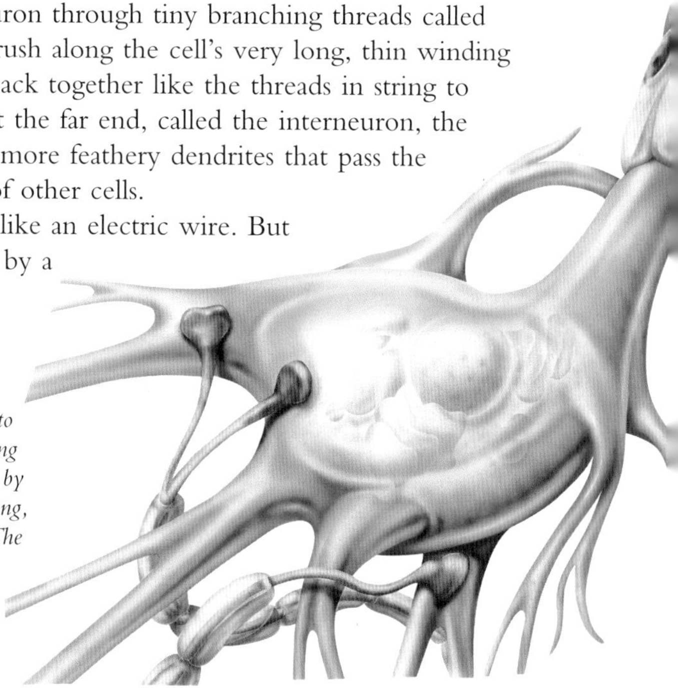

NERVE CELL
This illustration shows a nerve cell massively enlarged. Some nerves have to carry especially urgent messages over long distances. These nerves are surrounded by thick insulation, to keep the signal strong, just like the cable of a TV antenna. The insulation is called a myelin sheath, which is actually a series of long, flat cells called Schwann cells, wrapped like Swiss rolls around the axon.

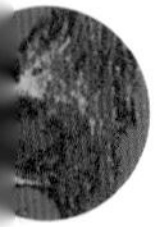

130

and chemistry. The secret lies in the electrical difference between the inside and outside of the nerve. On the outside is an excess of positively charged sodium ions (particles); on the inside is an excess of negatively charged potassium ions. Normally, the membrane (skin) of the cell stops sodium ions from drifting in and potassium ions from drifting out to balance out the charge. But when a nerve signal arrives, tiny gateways open in the membrane, letting the sodium atoms drift in. As sodium seeps into the cell, it opens up gateways farther along the nerve. This in turn lets in more sodium, which opens gateways even farther along the nerve, letting in more sodium. And so the signal is passed right along the nerve.

NERVE CELL
This illustration shows a nerve cell massively enlarged under a microscope.

Nerve facts
- Each neuron has only one axon, which can range from tiny to more than 3 feet (90cm) long.
- Myelinated nerves can transmit nerve signals 200 times faster than unmyelinated ones.
- The fastest nerve signals travel at more than 250 mph (400km/h).
- After passing a signal, a nerve recovers ready to pass another signal in less than one hundredth of a second.
- There are 100 billion neurons in your brain.
- Each neuron is linked to hundreds of others.
- The first person to see a neuron was the Italian scientist Camillo Golgi, who accidentally knocked a piece of owl's brain into silver nitrate—which stained the nerve cells and made them visible under a microscope.

NERVE GAPS

Nerve signals whiz from neuron to neuron all around the body. Yet no two neurons actually touch. Instead, there is a gap between them where they connect, called a synapse. For a nerve signal to be sent on, it must be transmitted across this gap by special chemicals released by the sending nerve, called neurotransmitters.

Neurotransmitters work a bit like keys in locks. Each receiving nerve has receptors that accept only certain neurotransmitters. To pass on the nerve message, the transmitter must be one that fits the nerve cell receptor in the same way that a key fits a lock. When the transmitter fits, it opens chemical gates in the membrane of the receiving cell to start off the electrical changes that transmit the nerve signal on through the cell.

1 As the nerve impulse arrives, vesicles drift to the gap and release their contents

2 Neurotransmitters flood across the gap

3 Neurotransmitters dock in matching receptor sites

4 As sodium gateways open, the newly excited nerve cell fires off a signal

Neurotransmitter

3

4

Nerve cell

Arriving nerve impulse

1

2

Vesicle storing drops of neurotransmitter

Nerve gap, or synapse

Nerve membrane

PASSING THE SIGNAL

Droplets of neurotransmitters are stored in the nerve end in little sacs called vesicles. When a nerve impulse reaches a nerve end, the vesicles drift to the synapse and spill out their contents into the gap. Once released, the neurotransmitters flood across the gap to the next nerve cell. Here they slot into receptor sites as a key fits in a lock. As the neurotransmitters dock, they change the chemistry of the nerve membrane, opening gateways to let in sodium ions. This makes the nerve excited and pass on the nerve impulse.

Nerve gaps
Excitement and inhibition

- If every nerve signal was passed on by the synapse, we would be overwhelmed by nerve signals. So at some synapses the receiving cell reacts to the neurotransmitter by passing on the signal, but others react by blocking it. This is called excitation and inhibition.
- Some tranquillizing drugs work by increasing inhibition. Some stimulants increase excitation.

Neurotransmitters

- There over 40 neurotransmitting chemicals.
- Acetylcholine is involved in making muscles contract.
- Noradrenaline helps control heartbeat and blood flow.
- Dopamine works in parts of the brain that control movement. Poor response to dopamine may cause Parkinson's disease.
- Endorphins are "neuropeptide" transmitters used by the brain to control pain.

SPINAL CORD

The central nervous system, or CNS, is the focus of the nervous system, and every nerve message starts and ends here. The top of the CNS is the brain, but running all the way down from it through the hollow center of the backbone is a thick bundle of nerve fibers called the spinal cord. It is through this motorway of nerves that signals are carried to and from the brain.

The spinal cord does not simply relay messages; it can work independently of the brain. When a message comes from the body, the spinal cord sends out a signal in response long before the signal reaches the brain.

The spinal cord is so important that if it is damaged in any way, the result can be devastating. Someone whose spinal cord is badly damaged may be paralyzed from the neck down. This is why the cord is protected inside a bath of "cerebrospinal fluid" (CSF), three membranes or "meninges" and the bone of the spine.

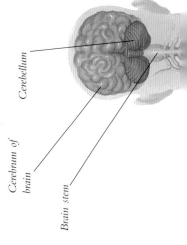

Cerebrum of brain

Cerebellum

Brain stem

SPINAL NERVES

Spinal nerves branch off in pairs on either side of the spine. There is one pair for each of the vertebrae, except for those of the very tail. So there are 31 altogether. They are grouped in the same way as the bones of the spine: cervical or neck; thoracic or chest; lumber or back; and sacral or lower back.

Cervial nerves (8 pairs) for neck, shoulders, and arms

Lumbar nerves (5 pairs) for lower back and thighs

Sacral nerves (6 pairs) for buttocks, genitals, legs, and feet

Nerves to arm

Thoracic nerves (12 pairs) for ribs, back, and abdomen

Filum terminale ends spinal cord

Spinal nerves

- The spinal cord is made of white matter, which itself is made of the long filaments of nerve cells, and gray matter, which is the main cell bodies.

- Nerve signals can only travel one way along a nerve, so some nerves carry signals down the spine (the descending pathway) and some carry them up (the ascending pathway).

- Your spinal cord reaches its full size—17 inches (43cm) long by 3/4 inch (2cm)—when you are four or five years old.

135

MOVING AND FEELING

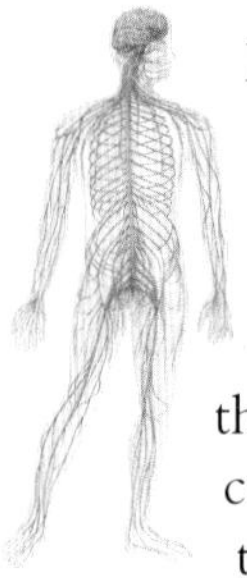

Besides the connecting nerves of your spine, you have two
kinds of nerve: nerves to make bits of your body move,
called motor neurons, and nerves to send signals from
your senses to your brain, called sensory neurons. Both
kinds of nerves are linked to two bands running around
the top of your brain, the motor cortex and the sensory
cortex. Every part of these two cortexes is linked by nerves
to a particular part of the body. Whenever you make a
move, your brain sends a signal out from the right part of
the motor cortex down the nerve to the right
muscle. Whenever you sense something, it is
because a nerve signal has traveled
from the sense to the right part
of the sensory cortex.

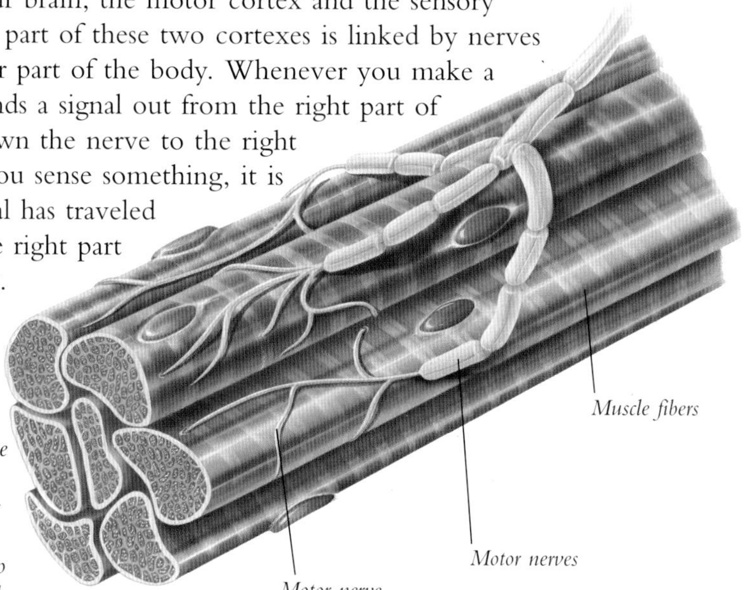

Muscle fibers

MUSCLE MOVEMENT
*Every major muscle in the
body has its own motor nerve
ending. Like a computer
telling a robot to move, your
brain sends out motor nerve
signals to the right muscles to
tell them to contract and make
a move, or to relax.*

Motor nerves

*Motor nerve
endings*

136

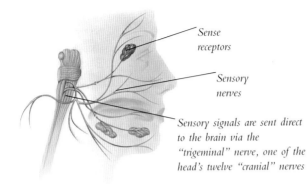

Sense receptors

Sensory nerves

Sensory signals are sent direct to the brain via the "trigeminal" nerve, one of the head's twelve "cranial" nerves

SENSITIVE FACE

Your face is packed with more sensory nerve endings than any other part of the body except for the hands. The lips in particular have a large number of nerve endings—and a correspondingly large area of your sensory cortex is devoted to your lips.

IN TOUCH

Whenever you feel a touch on your skin, you know exactly how hard you've been touched and where. You know how hard from the rate at which nerve signals reach your brain. You know where because each sensory nerve is linked to a different part of the cortex. You can identify the origin of each nerve signal by the bit of the cortex it stimulates.

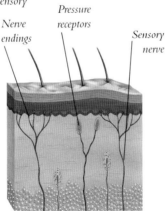

Nerve endings

Pressure receptors

Sensory nerve

Moving and feeling

Muscle control

- To send the right motor signals, your body has to know all the time just how tense or relaxed every muscle is. So each muscle has nerve receptors on the tendons to tell how tense it is.

- If the strain on a tendon goes up, signals are sent off to the CNS (see p134). The CNS then moderates motor signals going to the muscle to make sure the muscle contracts just the right amount.

Pathways

- Motor nerves cross over from one side of the body to the other at the top of the spinal cord. So signals from the right half of the brain go to muscles on the left of the body and vice versa.

- All the sensory and motor nerves pass through the central nervous system, except for the twelve "cranial" nerves of the head.

137

REFLEXES

Usually, motor and sensory nerves (p136) travel through the body side by side. But as they enter the spine, they split in two—and this gives the body one way of reacting to emergencies instantly. If you touch a hot plate, for instance, any delay in removing your hand could be damaging. The time it takes for a nerve signal to travel to the brain, your brain to react consciously, and send a signal back, could be long enough for a serious injury to occur. So the body has automatic, lightning-fast "reflexes" that happen so quickly you are aware of them only after they happen. Many work by short-circuiting the signal sent by the sensory nerve where it enters the spine. A link called an interneuron connects it directly to the motor nerve and fires it automatically.

NOT DROPPING OFF
Many reflexes are learned, not wired in from birth. Any activity demanding rapid physical reactions will involve reflexes that have been programmed into the body's nervous system by repeated experience. A climber's life may often depend on an instant reflex reaction.

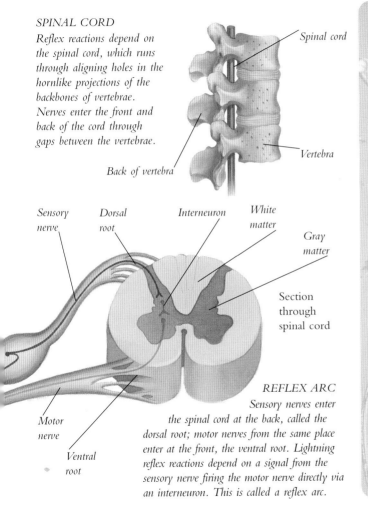

SPINAL CORD

Reflex reactions depend on the spinal cord, which runs through aligning holes in the hornlike projections of the backbones of vertebrae. Nerves enter the front and back of the cord through gaps between the vertebrae.

Spinal cord

Vertebra

Back of vertebra

Sensory nerve

Dorsal root

Interneuron

White matter

Gray matter

Section through spinal cord

Motor nerve

Ventral root

REFLEX ARC

Sensory nerves enter the spinal cord at the back, called the dorsal root; motor nerves from the same place enter at the front, the ventral root. Lightning reflex reactions depend on a signal from the sensory nerve firing the motor nerve directly via an interneuron. This is called a reflex arc.

Reflexes

Primitive reflexes

- When babies are born, they react automatically to certain things. Put something in their hand, and they grip it. When touched on the cheek with a finger, they turn and suck the finger. These are called primitive reflexes and last only a few months.

Inborn reflexes

- You are born with many reflexes, like shivering when cold and urination.
- The best known inborn reflex is the knee-jerk, which makes your knee jerk up when the tendon just below it is sharply tapped.

Conditioned reflexes

- Many reflexes are learned by experience, as certain pathways in the nervous system get reinforced by habitual use. This is called conditioning.
- It is a conditioned reflex that allows you to walk home along the familiar route without thinking about it.

UPPER BODY NERVES

The upper body is a mass of nerves radiating from the central nervous system—the brain and the spinal cord. Spreading out from the cord on either side are 31 nerves, which branch and interweave into hundreds of different nerves on their way to every part of the upper body, from the hands to the stomach. The hands have especially many nerve endings.

Higher up are the 12 nerves of the cranial system, the only major nerves that do not pass through the spinal cord, but go straight to the brain instead. Most run just inside—linking to major sense organs such as the eye (optic nerve) and the nose (olfactory nerve). But "vagus" means wanderer and the vagus links the throat, lungs, heart, and the digestive system.

OPTIC NERVE

The pictures picked up by the light-sensitive cells at the back of the eye are sent back to the brain along the optic nerve, the second of the twelve cranial nerves. This is not just a single nerve but a whole bundle.

Optic nerve

THE VAGUS

Running all the way from the brain stem to the stomach, the two vagus nerves, left and right, link the respiratory system, heart, and digestive system. The vagus nerves are involved in swallowing, coughing, sneezing, and speaking, as well as releasing the chemical acetyl choline to slow the heart, get stomach juices flowing, and make food move faster through the gut.

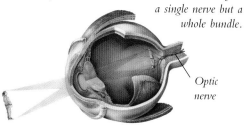

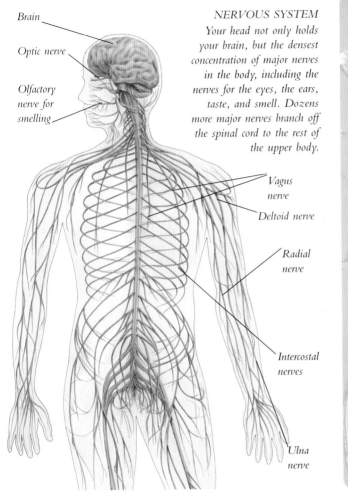

Brain

Optic nerve

Olfactory
nerve for
smelling

Vagus
nerve

Deltoid nerve

Radial
nerve

Intercostal
nerves

Ulna
nerve

NERVOUS SYSTEM

*Your head not only holds
your brain, but the densest
concentration of major nerves
in the body, including the
nerves for the eyes, the ears,
taste, and smell. Dozens
more major nerves branch off
the spinal cord to the rest of
the upper body.*

The cranial nerves

- *I* The olfactory nerve relays smells from the nose to the brain's smell center.
- *II* The optic nerve sends pictures from the eye to the brain's visual center.
- *III, IV & VI* The oculo-motor, trochlear, and abducent nerves control the eyes.
- *V* The trigeminal nerve relays sensations from around the eye, face, and teeth.
- *VII* The facial nerve stimulates taste buds, the ear, and salivary and tear glands.
- *VIII* The vestibulocochlear nerve relays sound and balance signals from the ears.
- *IX and XII* The glossopharyngeal and hypoglossal nerves are involved in swallowing and sensations from the throat.
- *X* The vagus or wanderer nerve controls many vital bodily functions including breathing, coughing, and digestion.
- *XI* The spinal accessory nerve helps you speak.

LOWER BODY NERVES

The main nerve in the lower body, and the largest nerve in the body, is the sciatic nerve, which runs down each leg. The sciatic nerve is the brain's link to the hip joints, the thigh muscles, and the skin on the back of the thigh.

The sciatic nerve is a branch of the "sacral" nerves that branch off the spinal cord at the very end, in the hips. When they leave the cord, nerves don't always branch and then stay forever apart. They often meet again and interweave, forming a tight bunch called a plexus. The sciatic nerve goes through the "sacral plexus" as it leaves the spinal cord, then splits behind the buttock and runs down the back of each thigh.

Above the back of the knee, the sciatic nerve splits again into two branches—the tibial and common peroneal nerves.

BALL JUGGLING
When a soccer player bounces a ball on his knee, his brain is sending a rapid stream of signals down his spinal cord and through the sciatic nerve to his thigh. Data streams back the same way.

142

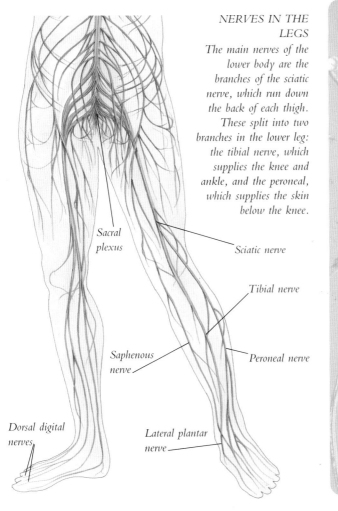

NERVES IN THE LEGS

The main nerves of the lower body are the branches of the sciatic nerve, which run down the back of each thigh. These split into two branches in the lower leg: the tibial nerve, which supplies the knee and ankle, and the peroneal, which supplies the skin below the knee.

Sacral plexus

Sciatic nerve

Tibial nerve

Saphenous nerve

Peroneal nerve

Dorsal digital nerves

Lateral plantar nerve

Leg nerve problems

Sciatica

- The most common problem with lower body nerves is "sciatica," which affects many old people. This is a problem with the sciatic nerve that happens when the disk of cartilage between two backbones collapses. The bones then squeeze the nerve's spinal root (see p139) and stop it from working.
- Sciatica can also be caused by a dislocated hip.
- At its worst, sciatica can cause numbness of skin in the legs, and paralysis below the knee.

Other nerve problems

- If you break your fibula—your calfbone—the peroneal nerve may be damaged. Your foot then hangs limply. This is known as foot-drop.
- Foot-drop can also be caused by inflammation of any of the nerves that supply the foot, which can be due to diabetes (see p168) or multiple sclerosis.

THE BRAIN

Your head encases the most complex structure in the entire universe, a human brain. Packed with an incredible network of nerve connections, a human brain can not only perform many tasks way beyond any of the world's fastest computers. It can also think for itself...and what's more, it knows it too.

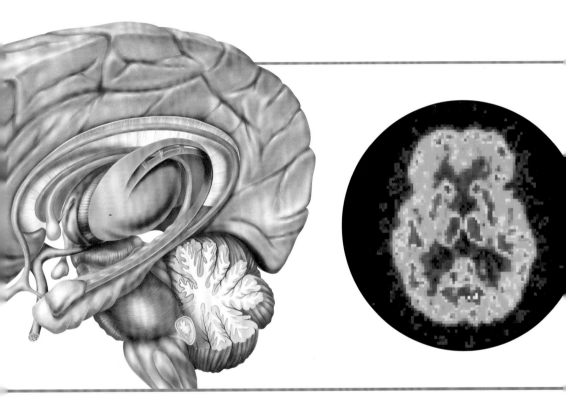

THE BRAIN

Inside your head you carry the most amazing structure in the universe: a human brain. It looks like little more than a large, soggy gray walnut with its wrinkled surface. But within this soft mass are billions of inter-linked nerve cells. The chemical and electrical impulses whizzing through all these cells create all your thoughts, record every sensation, and control

nearly all your actions. Every second of your life, your brain is receiving signals from the rest of your body and issuing instructions via the body's network of nerves. Even more amazingly, it lets you go on thinking, even when there are no inputs or outputs.

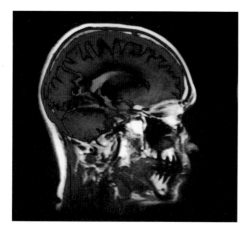

SEEING THE BRAIN
This model of a head with the top of the skull cut off (right) shows the brain's two halves or "hemispheres" and the wrinkles that cover its surface. But scientists can only see inside a working brain with the aid of special scanners, which reveal brain activity (left) with traces of chemicals injected into the blood.

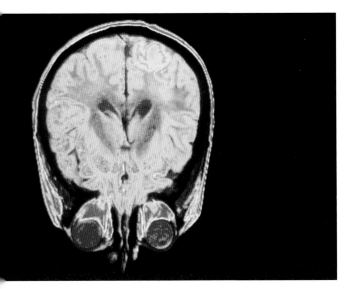

BRAIN SLICE
This scan shows a "slice" through a living brain from above in computer colors. The yellow mass is the brain, the blue the skull, and the red disks are the eyeballs.

VULNERABLE BRAIN

Even though it is no bigger than a bag of sugar, the brain's buzzing activity demands huge amounts of energy—and brain cells also depend on oxygen in the blood. If the blood supply to your brain were cut off, you would lose consciousness within ten seconds, and die within minutes.

Brain facts

- There are 100 billion cells in your brain, each connected to as many as 25,000 others—so there are 25,000 times 100 billion nerve connections in your brain, or two and a half million billion.
- Girls' brains weigh, on average, 2.5 percent of their body weight; boys' brains weigh 2 percent of their body weight.
- The brain makes up less than 3 percent of your body's weight, yet demands more than 25 percent of its blood supply.
- The membrane between brain cells and the blood supply is called the blood-brain barrier.
- The blood-brain barrier lets through only the finest particles, including glucose.
- The layout of the human brain and nervous system is much the same as other animals—but the top of the brain, the cerebrum, is much bigger.

AREAS OF THE BRAIN

All the brain looks much the same to start with, but a closer look shows that it actually has quite a varied structure. It is split into two halves or hemispheres, left and right, linked by a huge bundle of nerves called the corpus callosum. Surprisingly, the left half of your brain controls the right side of your body and vice versa.

Each side has three main regions. Deep in the center, connected to the spinal cord is the brain stem and the thalamus which sits on top of it. This is where basic functions like breathing and heart rate are controlled without your awareness. The cerebellum is a plum-sized

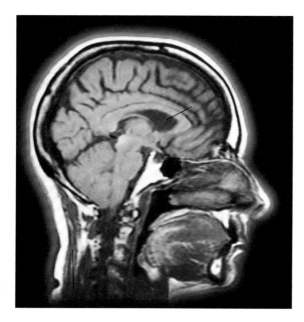

INSIDE THE HEAD
The human brain evolved from the middle out. So it is oldest, and most primitive—and most like other animals—right at the center. It is here that basic functions are controlled and emotions originate. The large cerebrum outside developed most recently. This is where our thoughts occur and is much bigger in humans than in other animals.

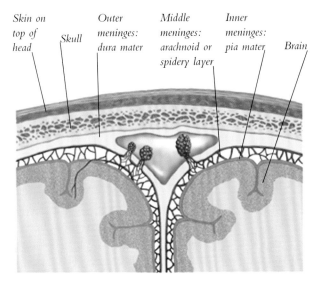

Skin on top of head | Skull | Outer meninges: dura mater | Middle meninges: arachnoid or spidery layer | Inner meninges: pia mater | Brain

TRIPLE LINING
The brain is so valuable that it is encased inside your bony skull and floats in a pool of "cerebrospinal fluid" which helps absorb shocks. Three layers of membranes called "meninges" are wrapped around it for extra protection.

lump growing from the back of the stem which controls balance and coordination. Most of the brain, though, is the cerebrum, which wraps around the thalamus like a big cherry around a pit. This is where you think, and complicated tasks such as memory, speech, and conscious control of movement go on.

Brain areas
- The outer surface of your brain, like the peel on an orange, is a layer called the cortex, just 0.16 inch (4mm) thick.
- The cortex is the place where all the messages from the senses are received and where all commands to the body to move come from.
- Some parts of the cortex co-ordinate these signals.
- The cortex is also called gray matter, because it is pinkish gray. Underneath the cortex, the nerves are sheathed in white myelin, and this is called white matter.
- The brain stem is made of the medulla (the stalk), with the pons and midbrain on top.
- The rounded ends of the cerebrum are called lobes. The upper lobes at the front, called the frontal lobes, are the really clever bits of the brain. The temporal lobes, the lower lobes, are linked

THE BRAIN INSIDE

The outside of the brain is a dense mass of brain cells where our thoughts occur. The middle of the brain is not so densely packed, but there are all kinds of different nerve centers, each with its own task. Most are designed to control different body activities automatically, and you are barely aware of their activity. The hypothalamus, for instance, monitors body heat and water and sends out signals to correct them when they are wrong. But these nerve centers can also send out signals to the rest of the brain and the body which have a great effect on how you behave. The hypothalamus, for instance, also helps wake you up and makes you more aggressive, while the limbic system is involved in many of your emotions.

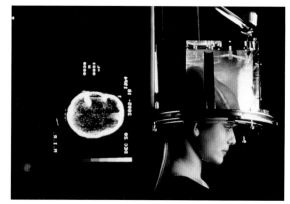

BRAIN SCAN
A variety of clever scanning techniques have enabled scientists to see inside living brains. They can actually see how certain areas become more active when a person is performing particular tasks. In this way, they have managed to identify areas of the brain associated with everything from memory to speaking and singing.

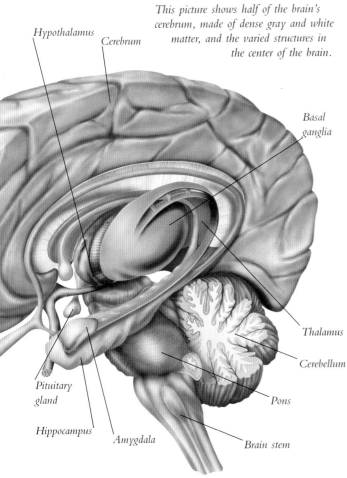

INSIDE THE BRAIN
This picture shows half of the brain's cerebrum, made of dense gray and white matter, and the varied structures in the center of the brain.

Hypothalamus

Cerebrum

Basal ganglia

Thalamus

Cerebellum

Pons

Pituitary gland

Hippocampus

Amygdala

Brain stem

Deep inside the brain

- The two egg-shaped thalamuses are relay stations for data on its way to the cerebrum.

- The hypothalamus controls body heat and water, and makes you feel hungry. It also makes you aggressive and wakes you up.

- The pituitary gland is a pea-sized ball just inside your head, behind your nose. It is the master hormone control gland (see p167).

- The limbic system is the ring of nerve centers around the thalamus. It not only processes smells, but emotions and memories too—which may be why smells can bring back memories so strongly.

- The hippocampus is a pair of small club-shaped nerve centers linked with moods, willpower, and learning.

- The amygdala seems to link moods—especially anger—with the body processes controlled by hormones.

THE MIND

Your mind is all your thoughts put together. Your brain is just the mass of nerve cells in your head. Many scientists say they are both the same thing, but just how the brain can create all your mind's thoughts is one of science's great mysteries.

What scientists know is that certain parts of the brain are linked to certain kinds of thought. At the back of the brain, for instance, on the "occipital lobe" is an area linked with sight. Farther forward is the place where signals from the ears are processed.

You decide what you're going to say—and understand what others say to you—in a part of the brain called Wernicke's area. But when you actually speak, your brain forms the words you are going to speak and issues instructions to your tongue in an area called Broca's area.

Motor cortex, where instructions for muscles to move are sent

Somato-sensory cortex, where sense signals are received

Frontal lobes deal with thoughts, creative ideas, and your personality

Broca's area forms words for speaking

Temporal lobe, where sounds are interpreted

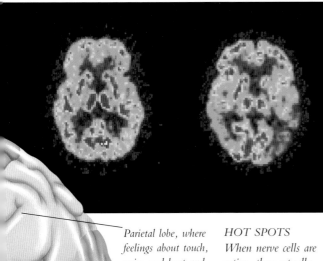

Parietal lobe, where feelings about touch, pain, and heat and cold register

Occipital lobe, where what your eyes see is interpreted

Wernicke's area is where we understand words

Cerebellum deals with balance and posture

HOT SPOTS

When nerve cells are active, they actually generate small amounts of radiation, which can be picked up on sensitive scanners. So scientists can see which area of the brain is active when someone is, say, reading.

BRAIN AREAS

The different colors on this picture of the brain show just some of the different areas of the brain linked to particular activities.

Association areas
- The brain's intelligent activities seem to go on in the outer layer, the cortex.
- Scientists locate things on the cortex according to different mounds or "lobes"—the frontal and temporal lobes at the front and the parietal and occipital lobes at the back.
- The frontal lobe is where your bright ideas occur. The motor cortex (see p137) is along the back.
- The temporal lobe is where you hear and understand people speaking.
- The parietal lobe is where touch, heat, cold, and pain are recorded. The somato-sensory cortex (see p137) is along the front.
- The occipital lobe is where what your eyes see is interpreted.
- Areas of the lobes that are not part of the motor and sensory cortexes or linked to sight and hearing are called association areas, because they may be associated with certain activities.

153

SLEEP

You will probably spend about a third of your life asleep. For most people at least six hours sleep every 24 hours is essential—although no one knows really quite why. Surprisingly, perhaps, most body activities—heartbeat, digestion, and so on—carry on just as normal while you're asleep. Even your brain does not shut down, receiving incoming signals all the time. However, as nerve cells in the brain fire, they send out minute electrical pulses that sweep across the brain. The pattern of these brain waves is very different when you are asleep to when you are awake.

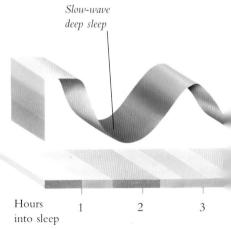

Slow-wave deep sleep

Hours into sleep

1   2   3

DREAMING SLEEP

When you are awake, the brain is buzzing with activity, with so many nerve cells firing off that there seems to be no pattern to their activity. But as you begin to feel drowsy, some of the cells begin to fire rhythmically together. Regular electrical pulses called alpha waves sweep across the brain every tenth of a second or so, accompanied by slower pulses called theta waves. While you are still drowsing, this pattern may easily be broken, but when you finally drop off, the pattern settles down.

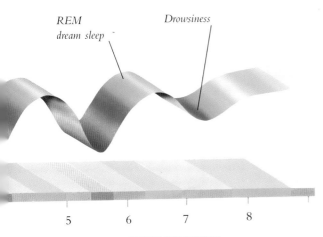

REM
dream sleep

Drowsiness

5   6   7   8

SLEEP RHYTHMS
As you sleep, REM sleep (when you dream)
alternates with periods of deeper sleep.

Through the night
Stages of sleep
- Stage 1: As you go to sleep, your breathing steadies and your brain waves begin to become steadier and more distinct.
- Stage 2: After a few minutes, stronger slow waves appear, along with bursts of rapid waves.
- Stage 3 and 4: Over the next 90 minutes, sleep gets deeper and slow waves become even stronger, until your whole brain seems to pulse every second or so.

Dreaming
- After 90 minutes or so of sleep, your brain suddenly begins to buzz with as much activity as when you are awake—yet you are very hard to rouse.
- Your eyes begin to flicker from side to side under their lids. This is called Rapid Eye Movement (REM) sleep and may be linked to dreaming.
- For the rest of the night REM sleep alternates with deeper, steadier sleep.

155

MEMORY

You may forget important things every now and then, but your brain actually has an amazing ability to remember things. You can remember thousands of words, for instance, and hundreds of faces. Scientists do not know exactly how memory works, but they think that the connections between brain cells are not fixed. Memories are stored as new connections are made. They are lost as the connections fail through lack of use.

Memories seem to be stored in three stages. First there is

"sensory" memory, when you go on seeing, hearing, or feeling something a little while after it stops. You can write your name in the air with a sparkler, for instance, because if you are quick you can still see the first letter as you draw the last.

Secondly, there is "short-term" memory. This stores things in the brain for a few seconds or minutes. When you look up someone's

MEMORY STORES

Sensations arrive in the brain via the limbic system, but the brain stores different kinds of memories in different ways. Memories of particular events, for instance, are formed in the hippocampus then stored in the cortex. Skills like violin playing and football, are stored in the cerebellum after repeated practise.

Cortex

Limbic system

Hippocampus

Cerebellum

KINDS OF MEMORY
Your long-term memories are of various kinds. "Episodic" memories are dramatic events like falling off a bike. A phone number is a "semantic" memory. "Nondeclarative" memories are skills you learn, like playing a

phone number and remember it long enough to dial, you are using short-term memory.

Thirdly, there is "long-term" memory, which can last all your life. This is when especially strong events get impressed on your brain or you go over things again and again until they are stuck in your brain.

Long memories
- You remember things for a long time in two main ways: declarative and non-declarative memories.

Declarative memories
- Declarative memories are of two kinds: semantic and episodic.
- Semantic memories are memories of facts such as words, phone numbers, dates and so on. The brain seems to store these in the temporal lobe (see p153) on the left of your brain.
- Episodic memories are memories of dramatic events. Here you don't just remember a fact, but all the things you felt and saw.
- Declarative memories are formed in the hippocampus and stored in the cortex.

Nondeclarative memories
- Nondeclarative memories are skills you have to teach your body by practicing them over and over again, like playing football or the violin. These are stored in the cerebellum.

COORDINATION

To make even a simple move, like picking up a notebook, the brain has to issue precise signals to a range of different muscles. To issue these signals, the brain must get constant feedback telling it exactly where each muscle is.

This feedback is provided by special sensory nerve endings in the muscles called proprioceptors—which can tell you where any part of your body is, even with your eyes closed. There are also other kinds of proprioceptors in tendons and muscles which tell your brain when the tendon or muscle is being stretched.

FAST SERVICE
Serving a tennis ball demands an amazingly complex coordination of muscles. As the ball comes down toward the racket, the eyes follow it closely. All the while there is a constant stream of sensory signals from the proprioceptors in the muscles of the racket arm and an equally constant stream of motor signals flowing back from the brain. The result is that even as the racket arm moves at an astonishing speed through the air—just a fraction of a second— your brain can make slight adjustments to make sure it hits the ball squarely.

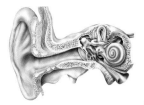

INNER EAR
Balance depends on a variety of receptors, but mainly upon the way liquid swills around the three semicircular canals of the inner ear—working like a 3D spirit level.

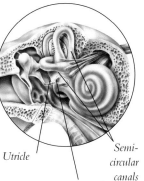

Utricle

Saccule

Semi-circular canals

PERFECT BALANCE
As this gymnast balances on the bar, the organs of her middle ear tell her exactly how upright she is, feeding data back to her cerebellum that sends signals to the muscles to adjust.

The balance organs
• The body's main balance organs are in each inner ear. They consist of the three tiny fluid-filled semicircular canals and a pair of spaces called the utricle and saccule.

• When your head nods or shakes, fluid washes through the canals. Since the fluid lags behind, hair receptors on the inside of the canals are bent back by the fluid and send signals to your brain.

• Your brain is told which way your head moves and how fast by the degree and direction the canal hairs bend.

• The utricle and saccule tell your brain if you tilt your head or body.

• Data from the balance organs is fed to the cerebellum (see p151), which also gets data about body's position from receptors in joints and places like the soles of your feet.

• The cerebellum processes all this data and coordinates muscle movement in conjunction with the cortex.

CHEMICAL CONTROLS

Coordinating all the systems of the body along with the nervous system is a remarkable array of chemical messengers. These messengers, called hormones, are little drops of chemicals released by glands, each of which has a particular effect on particular cells. If nerves are like the Internet, carrying messages instantly, hormones are like letters—slower and less direct, but longer-lasting.

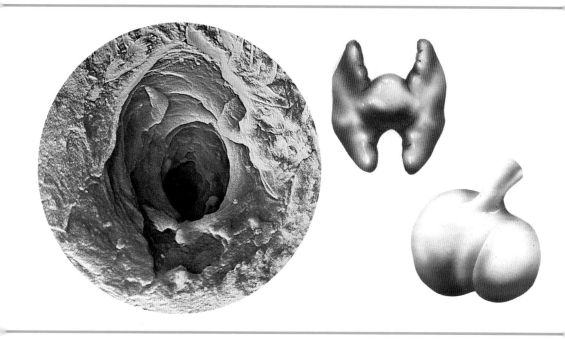

161

HORMONES

Working hand in hand with your nervous system in controlling the body are chemical messengers called hormones. Like nerves, they carry signals from one part of the body to another. But while nerves transmit messages on target instantly, hormones circulate in the bloodstream and provide a slower, longer lasting, often more widespread message. Nerves give the instant results you need, for example, when moving a muscle. Hormones give the more general controls to keep the body functioning properly.

The range of body processes controlled by hormones is enormous, and every single cell is influenced by hormones in some way or other. They are actually just simple chemical molecules, but each has a particular shape. A hormone's shape is not only its identity card but the message it carries, and they work by altering a cell's chemistry.

To be effective, each hormone must deliver its message to its "target"

Feedback

- Hormones are not controlled by the brain like nerves. They depend on working automatically. Hormones are released from their store only when they receive the right trigger.
- The trigger may be a chemical change in the blood, or it may be another hormone washing past.
- The release of many hormones is controlled by elaborate "feedback" mechanisms.
- The hormone insulin, for instance, tells the liver to stop releasing glucose into the bloodstream. If levels of glucose in the blood rise too high, the pancreas where insulin is made reacts by sending out more insulin. If levels of glucose in the blood drop too low, the pancreas sends out less insulin, so the liver releases more glucose.

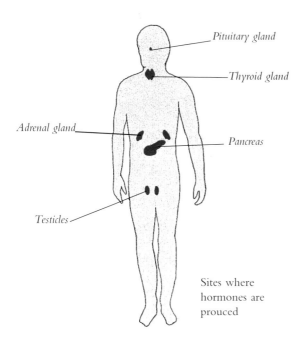

Pituitary gland

Thyroid gland

Adrenal gland

Pancreas

Testicles

Sites where hormones are prouced

Some hormones

Pituitary gland

- Growth hormone spurs growth and cell activity.
- TSH stimulates thyroid.
- ADH tells kidneys to cut urine production.
- FSH & LH stimulate ovaries.

Brain

- Endorphins and enkephalins reduce pain.

Thyroid

- Thyroid hormone raises cell activity.
- Calcitonin controls levels of calcium in the blood.

Adrenal glands

- Adrenalin and noradrenaline set off fight or flight response (see p164).

Pancreas

- Insulin and glucagon control blood sugar levels.

Ovaries

- Estrogen and progesterone control the menstrual cycle (see p222).

Testes

- Testosterone affects male sex organs.

cells, which might be either a particular kind of cell in just one place—ADH, for instance, targets just the collecting duct cells in the kidney—or it might be most body cells. Each target cell has receptor sites on its surface that the hormone locks onto if it is the right shape. So, as hormones wash past in the bloodstream, they find their places on cells like a key fitting into a lock.

ADRENALIN

Adrenalin is one of the hormones that gets us ready for action. Whenever danger threatens, the brain sends out an alarm signal to the adrenal glands on the top of the kidneys. At once, two hormones—adrenalin and noradrenalin—flood out from the adrenals into the bloodstream.

A third, cortisol, oozes out a little later.

What adrenalin and noradrenalin do is instantly prepare the body for emergencies. We can deal with most tricky situations by standing to face the threat or by running away. So their action is called the "fight or flight" response. They boost the supply of oxygen and energy-giving glucose to the muscles. They shut down distracting body processes like digestion. And they generally make us more alert.

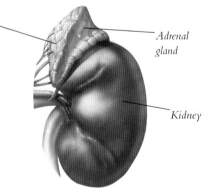

Adrenal gland

Kidney

ADRENAL GLANDS
The two adrenal glands are perched on top of the kidneys. Each has two distinct parts. It has a cortex, which, like the brain's cortex, is the outer layer. The medulla is the core. Cortisol comes from the cortex of the gland; adrenalin and noradrenaline are secreted by the medulla.

ADRENALIN RUSH
The thrill of snowboarding or skydiving brings a rush of adrenalin and noradrenalin. As they hit the blood, the heart beats stronger and faster to boost blood supply. Blood vessels widen and channel blood to the muscles. Eyes widen so you can see better, hair stands on end and skin goes pale and cold as blood to the skin is reduced. You even start to sweat as the body gets ready for the big heat.

Action hormones

Adrenalin and noradrenalin effects:
• heart beats stronger and faster.
• blood supply to muscles increases.
• blood supply to skin lessens.
• quicker, deeper breathing to boost oxygen to muscles.
• the eye's pupils widen.
• the skin goes pale.
• you sweat.

Cortisol effects:
• prepares your body for after-effects of danger.
• unlocks energy from fat.
• mobilizes amino acids to help repair damaged cells.
• helps reduce pain—which is why people may feel no pain from a severe injury until long after.

Medical uses
• Adrenalin may be injected to restart a heart when it stops in a cardiac arrest. It may be used in acute asthma attacks.
• Cortisol is mimicked by corticosteroid drugs, which have a huge range of uses, for everything from asthma to arthritis.

THYROID AND PITUITARY

The thyroid gland and the pituitary gland are no bigger than cherries. But these two glands have a dramatic effect on the way your body works.

The pituitary is situated in the middle of your brain, just behind your nose. It is sometimes called the master gland, because it is the gland that sends out the hormones that direct the operation of all the body's other endocrine glands (see p168).

The thyroid is situated on the front of your throat, just below your voice box. This gland secretes three hormones—thyroxine (T4), triiodothyronine (T3), and calcitonin. T3 and T4 are the hormones that control how energetic you are by stimulating cells to burn more glucose or less.

If your thyroid does not send out enough T3 and T4, you get very tired and grow fat, your skin gets dry and you get cold easily. If it sends out too much, you may get thin, nervous, sweaty, and overheated. The panel on the right shows how levels are kept just right.

GROWING UP
By setting the rate at which cells work, the thyroid plays a major part in making sure you grow properly. The pituitary makes growth hormone, which makes cells grow bigger and multiply faster, so plays an equally important part in making sure you grow normally.

THYROID CONTROL
Levels of the thyroid hormones T3 and T4 in the blood are kept just right by continual feedback.

Thyroid

Pituitary

When levels of T3 and T4 in the blood rise, it dulls the reaction of the pituitary to the hormone TRH	When levels of T3 and T4 in the blood drop, the hypothalamus makes more TRH.
↓	↓
Hypothalamus oozes the same TRH	Hypothalamus oozes extra TRH
↓	↓
Dulled pituitary secretes less TSH	Extra TRH makes the pituitary secrete more TSH
↓	↓
Less TSH means thyroid makes less T3 and T4	Extra TSH means thyroid makes more T3 and T4
↓	↓
Levels of T3 and T4 in blood drop to normal	Levels of T3 and T4 in blood climb to normal

Metabolism and the thyroid

- Metabolism is all the chemical processes going on in your body—especially cellular conversion of glucose into energy. Metabolic rate is the rate processes work.
- People have naturally different metabolic rates. It goes up when you are working hard or are frightened.
- Your metabolic rate is normally regulated by the thyroid hormones T3 and T4 circulating in the blood.
- Just how much T3 and T4 the thyroid releases into the blood depends on how much thyroid stimulating hormone (TSH) it gets from the pituitary.
- The amount of TSH the pituitary sends out depends, in turn, on how sensitive it is to the hormone TRH made by the hypothalamus (p150).
- If levels of T3 or T4 drop, the pituitary sends out extra TSH to get the thyroid to make more.
- If they rise, the pituitary is less sensitive to TRH and sends out less TSH.

ENDOCRINE GLANDS

Many of the body's hormones are made in special glands right next to major blood vessels. These glands are called the endocrine glands—which is why hormone control is often known as the endocrine system. "Exocrine" glands, such as the salivary glands, secrete their juices into organs or outside the body and work only locally. Endocrine glands secrete them directly into the bloodstream to be rushed around the body to different organs and tissues. Each endocrine gland secretes a particular set of hormones, and so controls a particular set of body processes. But many are under control of hormones from the pituitary gland.

OVARIES

A woman's ovaries make the hormones estrogen and progesterone. These hormones affect a huge number of female bodily activities, including menstruation and ovulation (see p222).

TESTES

A man's testes make the hormone testosterone, which is involved in a number of male bodily activities, including sperm production and creating male characteristics, like a deep voice.

ADRENALS

The adrenals are the glands on top of the kidneys that release the hormones that prepare the body for emergencies—adrenalin, noradrenalin, and cortisol. This is the "fight or flight" response (see p164).

PANCREAS

The pancreas secretes the hormones insulin and glucagon. These control the amount of glucose in the blood by making the liver either store it as glycogen, or release it.

THYROID

The thyroid glands in the neck produce thyroid hormones such as thyroxine, triodothyronine, and calcitonin. These keep body systems alert, controlling how fast cells burn food, body heat, and bone growth. The parathyroid glands secrete parathyroid hormone, which controls calcium in blood.

PITUITARY

The pituitary is the master control gland of the endocrine system. It secretes hormones that trigger the adrenals, the thyroid, the testes, and the ovaries into action.

Endocrine hormones

Estrogen and progesterone

- Estrogen and progesterone are the controllers of the reproductive system in women and girls.
- Estrogen begins the cycle of egg production; progesterone ends it.
- Drugs based on these hormones are used in contraceptives.

Testosterone

- Testosterone is the most important of the male hormones, called androgen hormones.
- Male hormones boost muscle growth and androgen drugs are illegally—and dangerously—used by body builders.

Cortisol and others

- Cortisol helps the body recover from stress, and is the basis of important drugs.
- Aldosterone controls the body's salt levels.
- Insulin maintains blood sugar.
- Glucagon controls blood sugar.

BODY HEAT

Sometimes it is hot; sometimes it is cold. But whatever the weather, your body stays at the same temperature, 98.6°F (37°C). The body needs this steady temperature to ensure body reactions work well, and achieves it in a number of ways.

You keep warm mainly by eating. Indeed nearly all the food you eat is turned into heat. Some is made by the workings of organs, but most comes from muscle activity. When working hard, muscles can make as much heat as an electric fire. You keep cool by breathing—as long as the air is cooler than body temperature. Your body also loses heat through the skin and by sweating.

PERFECTLY WARM
Your body is equipped with a range of automatic ways of heating and cooling.

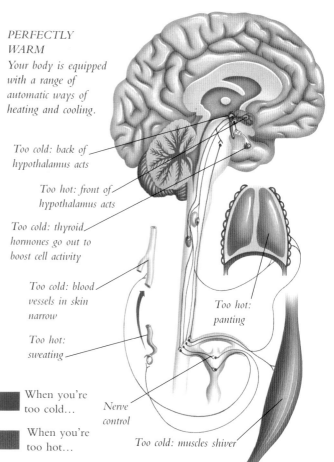

Too cold: back of hypothalamus acts

Too hot: front of hypothalamus acts

Too cold: thyroid hormones go out to boost cell activity

Too cold: blood vessels in skin narrow

Too hot: sweating

Too hot: panting

When you're too cold...

When you're too hot...

Nerve control

Too cold: muscles shiver

170

SCORCHING RETURN
When playing vigorous sport, the body has to lose some of the heat generated by muscle activity by sweating through pores on the skin (greatly magnified on right).

TOO HOT OR TOO COLD

With the air about 56°F (12°C), the heat you gain by heating roughly balances the heat you lose through the skin and by breathing. But sometimes the temperature is much higher or lower. Or your body may generate extra heat by violent exercise. Then the body's temperature control—the hypothalamus in the brain (see p150)—uses the tools illustrated left.

Heat controls
The body's thermostat

- The hypothalamus checks body heat with sensors in the skin, in the core of the body and in the hypothalamus itself, taking the temperature of the blood flowing past.
- The front of the hypothalamus issues instructions if it is too hot; the back if it is too cold.

Staying cool

- You may pant.
- Surface blood vessels widen to carry more heat away through the skin.
- You sweat more. Sweat not only takes warm water out, but as the moisture evaporates it actually cools your skin.

Getting hot

- You shiver to generate muscle heat and cells work faster.
- Surface blood vessels narrow to cut down heat loss through the skin.
- Skin hairs are erect, giving goose bumps. This may be a hangover from ancestors who had hairier bodies—so raising hairs trapped warm air next to the skin.

171

THE SENSES

The senses tell you about the world around you—its sights, its sounds, its heat, its cold, and much more. Even when you're asleep, your body is alive to sensations, both from inside and outside the body. There are five major senses—sight, hearing, smell, taste, and touch—and your brain receives a constant stream of nerve signals from each of these senses, keeping your brain up to date on what is going on.

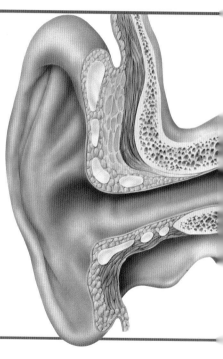

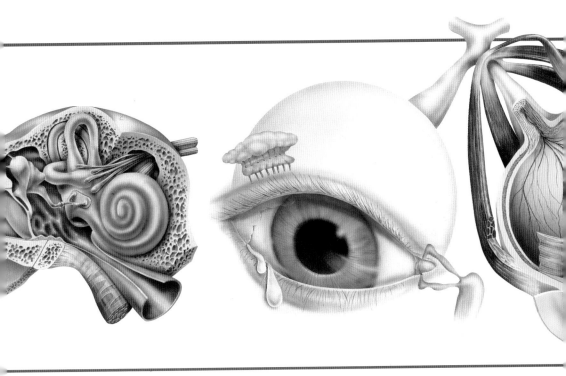

THE EYES

The picture formed on the back of your eye is a few millimeters across, yet it seems so big that it hardly occurs to you that it is just a picture, like a camera picture.

Your eyes combine the optical quality of the best cameras with astonishing versatility. A good camera may match the eyes for sharpness, but none can focus on both a speck of dust and distant galaxies. What's more, your eye works in both starlight and sunlight, a difference in brightness of a hundred billion times. It does this by adjusting its sensitivity to suit the light—which is what happens when your eyes gradually become accustomed to a dark room.

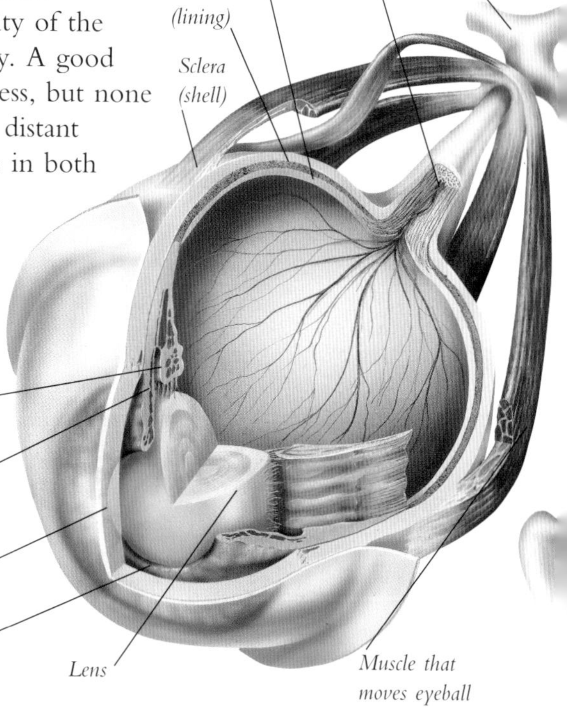

Retina (light-sensitive layer)

Optic nerve

Optic chiasma

Choroid (lining)

Sclera (shell)

Muscle that adjusts iris

Suspensary ligaments holding lens

Cornea

Iris

Lens

Muscle that moves eyeball

174

RODS AND CONES

The retina at the back of the eye is made of millions of light-sensitive cells called rods and cones, which send signals to the brain along the optic nerve ("optic" means for eyes). The process is described in the panel.

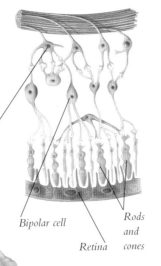

Ganglion cell

Bipolar cell

Rods and cones

Retina

THE EYES

Your eyes are two tough little balls filled with a jellylike substance called vitreous humor. Each is a bit like a video camera with a lens called a cornea at the front—the dark circle of your eye. This projects a picture on the back of your eye, called the retina. There is another lens behind the cornea, but this simply adjusts the focus of the cornea's picture.

Pupil

Tear duct

From eye to brain

- Whenever hit by enough light, each of the rods and cones in a little group sends a tiny electrical signal to their group leader, the bipolar cell.

- Instantly, the bipolar cell passes the message on to a ganglion cell.

- The ganglion is always firing signals down to the optic nerve. But the signals from the bipolar excite it to fire faster.

- Signals from all the ganglion cells in each eye whiz away down the divided highway of your two optic nerves. They meet at a crossroads in the middle of your brain called the optic chiasma.

- At the optic chiasma, one half of each "lane" goes off to the right of the brain and the other to the left, so each half of the brain gets half the picture from each eye.

[The journey of the picture is continued on page 177]

175

EYESIGHT

Your eyes work by focusing rays of light from whatever you look at onto the back of the eye. When you look at something—a flower, say—what you actually see is little chunks of light called photons bouncing off the flower towards your eyes. Streams of photons form rays of light that zoom into your eye through the cornea. Because the cornea is slightly denser than air, it bends or "refracts" the light rays so that they are steered closer together—so close together that by the time they hit the back of your eye and create a picture on the retina, the picture of the flower is just a millimeter high. This is called focusing. The eye's lens changes shape to bend the light rays more or less, according to how far away the flower—or any other object—is.

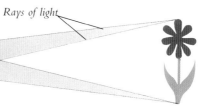

Rays of light

Cornea refracts rays

Lens adjusts degree of focusing to suit distance

Picture on the retina— upside down and tiny

UPSIDE DOWN AND TINY

As rays of light pass through the cornea, they are not only brought together to a focus; they also cross over. So the picture of the flower in your eye is not only tiny, but flipped upside down, too. Your brain sees this as normal.

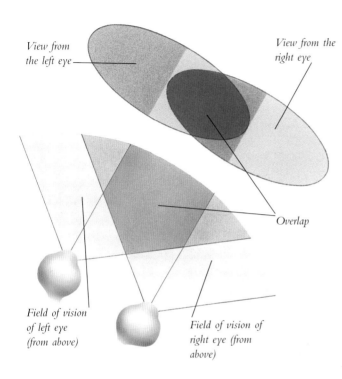

View from the left eye

View from the right eye

Overlap

Field of vision of left eye (from above)

Field of vision of right eye (from above)

STEREO VISION

Each eye gives us a slightly different view of the same object. The nearer the object, the greater the difference between the views. These slight differences combine in the brain to give us an impression of 3D depth and solidity. In the diagram above, the different views from each eye, and the area of overlap is greatly exaggerated. Only when things are very, very close to you is the difference so marked.

From eye to brain
[Continued from p175]

- From the optic chiasma signals go to the visual sorting office, the LGN, in the middle of your brain.
- The LGN splits the picture signal into different kinds of picture—pictures showing movement, pictures showing lines, pictures showing dark and light, and so on.
- The LGN sends sorted signals on down nerves called optic radiations.
- Next to the LGN is your brain's visual troubleshooter, the superior colliculus.
- The superior colliculus monitors pictures for anything that might mean danger. If it spots a danger—like a brick about to land on your head—it alerts the brain to action.
- All the pictures sent by the LGN arrive at the different screen's of your brain's multiplex cinema, the visual cortex. This is where you see what's going on, a fraction of a second after it hits the eye.

SEEING IN COLOR

You can see all the colors of the rainbow. But no one knows just how. Your eye is equipped with two kinds of cells: rods and cones. Rods can detect anything from very bright light to very dark light—but they cannot tell one color from another. Cones are the cells that give you color vision. It seems there are three kinds of cones.

Some are most sensitive to red light, some to green and others to blue. So once scientists thought we must see in terms of just three colors. They believed each of the rainbow of colors you see is created in the brain, depending on just how strongly each of the three kinds of cone are stimulated. But now they know it is a little more complex (see panel right).

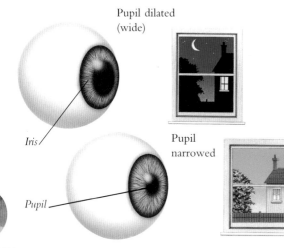

Pupil dilated (wide)

Iris

Pupil narrowed

Pupil

NIGHT VISION

In order to see well in dim light, the iris of your eye opens the pupil much wider to let in more light. In the dark, your pupil may become up to 16 times bigger. The sensitivity of the cells of the retina may increase, too, by over 100,000 times. This is called "dark adaptation"—and is why you are dazzled if someone flashes a bright light into your dark-adapted eyes. But only the rods adapt to the dark. The cones that give you color vision are just not sensitive enough to work in very dim light—which is why everything looks grayer in the dusk.

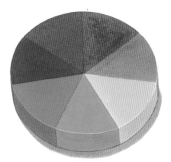

RAINBOW MIX

White light like daylight is a mix of every different color. Spin this wheel very fast so all the colors mix and you would see it as white. But all these colors can be made, too, by mixing just three basic colors of light—red, green, and blue.

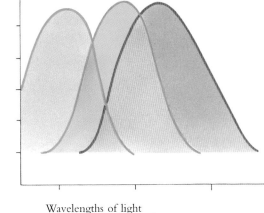

How strongly the cones react to each color of light

Wavelengths of light

COLOR REACTION

This graph shows how cones react to different wavelengths of light. It shows that there are three peaks—blue, green, and red. So there are three kinds of cone—blue-sensitive, green-sensitive, and red-sensitive.

Color vision

The trichromatic theory

- The three-color or "trichromatic" theory suggested you see all colors as mixes of three colors—red, green, and blue.

- This doesn't quite work because it doesn't show how we see colors like brown, silver, and gold.

The opponent-process theory

- This theory suggests we see colors in two "opposing" pairs—blue and yellow; red and green.

- When there is lots of blue light, yellow receptors in the eye react less (and vice versa).

- When there is lots of red light, green receptors react less (and vice versa).

Invisible colors

- Colors are different wavelengths of light. The shortest our eyes can see are violet; the longest are red. Some animals can see ultra-violet, light slightly shorter than violet. Others can see infra-red, light slightly longer than red.

OPTICAL ILLUSIONS

In some ways, each of your eyes is like a television camera, continually recording the world. But you don't see quite as a TV camera sees, for your brain is continually adjusting the picture, highlighting some bits and playing down others as it tries to make sense of it all. Your brain doesn't simply receive the picture but tries to work out what is a person, what is a car and so on. Without this continual interpretation, you could hardly get along in the world—but it does mean that every now and then, your eyes seem to play tricks on you. These tricks are called optical illusions.

OLD AND YOUNG

What do see when you look at this drawing? An old woman or a young one? Both are here. There is a young lady, with her face turned away from you. There is also an old woman—her mouth is the girl's neckband and her left eye is the girl's ear. Your brain tries to fit what it sees into a particular category. When there are various possible answers your brain will try to choose one. So you can't see both pictures at the same time. Once your brain has recognized both pictures, it continually switches between the two.

VASE OR FACES?

Even simple lines are enough to say to your brain "face" or "vase." So when you see a set of lines that could be either, your brain sees both.

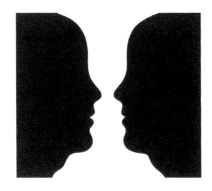

HOW MANY TRIANGLES?

There are actually none, but your brain takes the clue and fills in the gap. So most people will see a white triangle laid over the top of a pink one.

IMPOSSIBLE PRONGS

The visual clues on the right of this picture tell your brain this is a shape with three prongs. The clues on the left show a U-shape. Which is it?

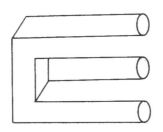

Tricks of the eye

Persistence of vision

- There is a limit to how fast your eye can take in changes. If a light flashes on and off faster than 50 times a second, your eye simply stays "on" and sees only a steady light, not a flashing one. This is called persistence of vision.

- By flashing still pictures up so quickly your eye can't see them changing, an illusion of movement can be given. This is how TV, film, and animation work.

Going backward

- Your brain is especially good at noticing movements, and is wired to pick up changes in the spread of light or in its brightness.

- This may create illusions. Stare at a spinning bicycle wheel long enough and you will see the spokes seem to move backward. Stare hard at water pouring into a bathtub from a faucet. The water soon seems to flow upward back into the faucet.

181

THE EAR

Sounds are just tiny vibrations in the air, and your ears are incredibly sensitive devices for picking up these tiny vibrations.

The flap of skin on the side of your head is only the entrance to the real ear, and simply funnels the vibrating air towards sensitive pressure detectors inside the head. The ear actually has three sections. The outer ear is the earflap and the ear canal, the tunnel into your head. Inside your head, in the "middle ear," the sound hits a taut wall of skin called the eardrum, shaking it rapidly.

As it shakes, it rattles three tiny bones, or ossicles. Even further inside, in the "inner ear," is a curly tube full of fluid, called the cochlea. As the ossicles vibrate, they knock against this tube, making waves in the fluid. Minute hairs waggle in the waves, sending signals along nerves to the brain.

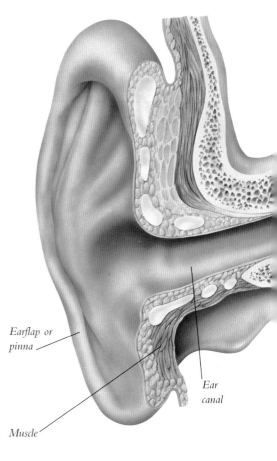

Earflap or pinna

Ear canal

Muscle

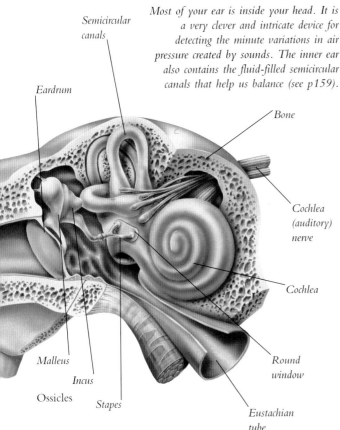

THE EAR

Most of your ear is inside your head. It is a very clever and intricate device for detecting the minute variations in air pressure created by sounds. The inner ear also contains the fluid-filled semicircular canals that help us balance (see p159).

Semicircular canals

Eardrum

Bone

Cochlea (auditory) nerve

Cochlea

Round window

Malleus

Incus

Ossicles

Stapes

Eustachian tube

Hearing

- Because you have two ears, you can pinpoint where a sound came from, though not always accurately.
- You can pinpoint sound because a sound to the left of you is slightly louder in the left ear than in the right ear and vice versa.
- If your hearing is normal, you can hear sounds as deep as 20 Hz, deeper than a bass drum.
- If your hearing is normal, you can hear sounds as high as 20,000 Hz.
- Sound intensity is measured in decibels on a geometric scale—that is, three decibels are twice as loud as two, and four twice as loud as three.
- If your hearing is normal, you can hear sounds as quiet as 10db (decibels), which is quieter than leaves rustling in trees.
- If your hearing is normal, you can hear sounds as loud as 140db or more, which is as loud as a jet engine close up.

183

INSIDE THE EAR

The inside of the ear is a very delicate and intricate structure—as it needs to be to pick up the tiny movements of the air that make up each sound.

To register clearly, the air movement has to be made bigger and clearer, and the three tiny bones of the ossicle—the hammer, anvil, and stirrup—are a kind of mini–amplifier. Whenever a sound vibrates the eardrum, the drum rattles the hammer against the anvil, and the anvil shakes the stirrup.

Because the hammer is biggest of the ossicles, it moves a long way with each vibration. The stirrup is smallest and vibrates only a little way, but each vibration is that much stronger. This extra force makes the vibrations strong enough to vibrate the fluid in the inner ear.

THE OSSICLE

The three bones of the ossicle are all known by Latin names which have simple, descriptive English equivalents. There is the malleus or hammer, the incus or anvil, and the stapes or stirrup. The hammer bangs against the anvil like a blacksmith's hammer on his anvil. The stirrup gets its name because it is shaped like a horse stirrup.

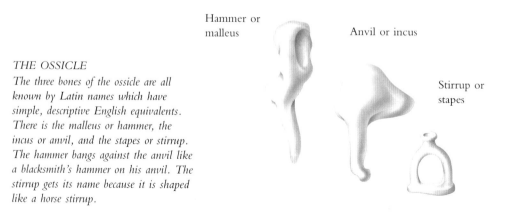

Hammer or malleus

Anvil or incus

Stirrup or stapes

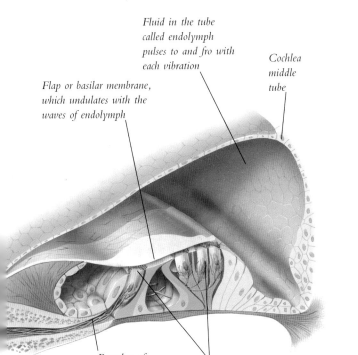

Fluid in the tube called endolymph pulses to and fro with each vibration

Cochlea middle tube

Flap or basilar membrane, which undulates with the waves of endolymph

Branches of cochlea nerve to brain

Hairlike receptors of organ of Corti

THE ORGAN OF CORTI

Vibrations are amplified by the time they reach the inner ear or cochlea, but they are still very tiny. So the cochlea contains a remarkably delicate device called the organ of Corti to detect every slight vibration (see panel).

Inner ear

The oval window

- The amplification begun by the ossicle is completed as the stirrup rattles on the membrane entrance to the inner ear, the oval window.

- The oval window is 30 times smaller than the eardrum, so the vibration is compressed and intensified.

The organ of Corti

- There are three tubes running through the cochlea. Inside the middle tube are rows of fine hairs under a membrane flap. These rows are the organ of Corti.

- When the stirrup knocks on the oval window, it sends pressure waves shooting around the cochlea. As they wash up and down, they wobble the organ's flap.

- When the organ's flap moves, it tugs on the hairs to and fro, playing on them like hands gliding over harp strings.

- As the organ's hairs are tickled, they send signals to the brain, revealing the sound.

TOUCH

Touch is the most widely spread of all your senses, with receptors all over your body, from head to toe. Some places have many receptors, such as your hands and face. Others, like the small of the back, have comparatively few.

The touch receptors in the skin react to four main kinds of sensation: a light touch, continuous pressure, heat and cold, and pain. All four of these sensations are felt in skin areas where there are only free nerve endings as receptors. But in some places there are a handful of different kinds of specialized receptors too, and each seems to respond to one kind of sensation more strongly.

There are five main specialized receptors, each named after its discoverer. Two respond mostly to a sudden touch or knock: Pacini's corpuscles and Meissner's endings. The other three respond mostly to steady pressure: Krause's bulbs, Merkel's discs, and Ruffini's endings.

When a receptor in the skin is stimulated, it fires nerve signals to the brain. The rate the nerve fires signals tell the brain, for instance, how heavy

FINGER READING
In many blind people, sensitivity to touch is so highly developed that they can read with their fingers. The Braille system gives letters as different patterns of raised dots. Blind people can even operate computers using braille.

186

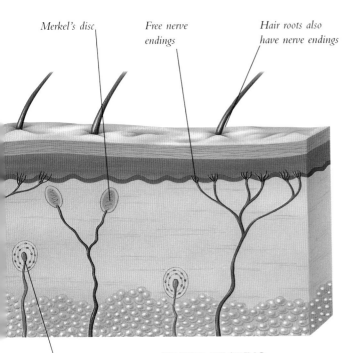

Merkel's disc

Free nerve endings

Hair roots also have nerve endings

Pacini's corpuscle

FINGER READING
This section through the skin shows some of the different types of touch receptor.

the touch is or how cold it is.

But the receptor does not go on firing forever. Instead, the firing rate falls off as it adapts to the stimulus. Once it has alerted the brain, it need only send an occasional reminder—which is why you cease to feel clothes

Special touch receptors

Sudden pressure

- Onion-shaped Pacini's corpuscles are the biggest and deepest of the touch receptors. Pacini's corpuscles respond to very heavy pressure and fast vibrations.
- Meissner's endings are little capsules just below the surface of the skin, especially on the palm side of the fingers. They detect light touch and vibrations and may be what enable you to distinguish delicate textures by feel alone.

Steady pressure

- Merkel's discs are bowl-shaped receptors that respond to continuous pressure.
- Krause's bulbs are bulb-shaped receptors deeper in the skin that are believed to respond to cold as well as steady pressure.
- Ruffini's endings probably respond to changing temperature as well as continuous pressure.

SMELL

The human nose is not quite so sensitive as that of some animals. But even so, your nose detects smells just by picking up a few airborne molecules of a substance, and it can distinguish more than 3,000 different chemicals.

Our sense of smell relies on a small patch inside the top of the nose called the olfactory epithelium. Although just 2.5 inches (6cm) square, it is packed with 10 million olfactory (smell) receptors. These receptors feed into a small area called the olfactory bulb, which then relays messages to the brain.

SENSE CENTER
The nose is divided into two halves by a wall of cartilage called the nasal septum. But both nostrils lead smells up to the olfactory patch high inside the nasal cavity. Here they stimulate tiny hairlike cells called cilia on the top of millions of olfactory receptor cells.

THE SCENT OF TULIPS
When you smell the sweet scent of flowers, your nose is actually reacting to small molecules given off by the flowers and wafted in the air up your nostrils.

PICKING UP A SCENT

On the top of each olfactory receptor are 20 or so tiny hairlike "cilia," and it is these that are stimulated by molecules of a smelly substance. For a smell to be detected, it must vaporize so that it can be wafted up the nostrils. It must then dissolve in water so it can get through the thick coat of mucus covering the cilia. The number of receptors stimulated tells you just how strong a smell is. The position of the cilia stimulated may be what helps your brain identify the smell.

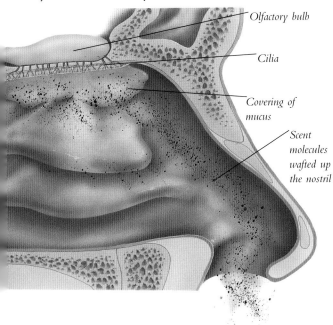

Olfactory bulb

Cilia

Covering of mucus

Scent molecules wafted up the nostril

Smell

- Nasal receptors react more to change than to steady stimuli. The nasal receptors may soon stop responding to a smell so that you have to sniff hard to go on smelling a faint scent.
- Most of the chemicals you can smell contain at least three atoms of carbon.
- The sense of smell is strongest in babies, and helps a baby recognize its mother.
- By the age of 20, your sense of smell will have dropped off by 20 percent.
- By the age of 60, your sense of smell will have dropped off by 60 percent.
- The smell of chemical "pheromones" in sweat may play a part in the attraction between men and women.
- Dogs have olfactory patches 30 times as big as humans.
- Humans can detect smells in concentrations of 1 part in a billion. Dogs are 10,000 times more sensitive.

189

TASTE

Our sense of taste is in some ways much vaguer than all the other senses, and we can only taste the difference between sweet food, salty food, sour food, and bitter food. However, when we eat we bring a whole range of other sensations into play—heat and cold, texture, look, and especially smell. So most of us can tell the difference between a huge range of foods, and trained wine and food tasters can detect and appreciate small variations in taste.

The body's taste receptors are the taste buds of the tongue, set in tiny pits. To reach them, food must first be dissolved in saliva, which is why you can't at first taste food that doesn't dissolve quickly. There are four kinds of bud, each responding to a different taste: sweet, salty, sour, or bitter. When any is set off, it sends a message to the brain.

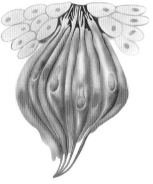

CHEESY FLAVORS
Our taste buds can detect
very little real difference
between all these cheeses.
But other senses such as
smell combine with taste
to reveal the rich
differences in flavor that
makes certain cheeses
so popular.

TASTE BUD
In each taste bud there is a cluster
of cells with tiny hairs on the end.
These hairs are washed over by
saliva pouring into the bud. If the
right taste is in the saliva, the hair
triggers off the receptor cell below.

190

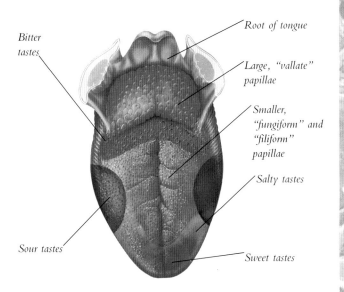

Bitter tastes

Root of tongue

Large, "vallate" papillae

Smaller, "fungiform" and "filiform" papillae

Salty tastes

Sour tastes

Sweet tastes

THE FOUR TASTES

The taste buds are located inside small bumps on the surface of the tongue called papillae. Taste buds that respond to sweet tastes are concentrated at the tip of the tongue. Salty flavors are detected most strongly just behind on the sides of the tongue. Sour tastes that make you wince produce the strongest response on the sides of the tongue, farther back still. Really bitter tastes set off a cluster of taste buds right at the back of the tongue, just above the entrance to the throat.

Taste

- There are 10,000 or so taste buds in various places over the tongue.
- In children and adults, there are no taste buds in the center of the tongue.
- A baby has taste buds all over the inside of its mouth, not just on its tongue.
- As you get older and older, your taste buds die off, and by the time you are 70 your sense of taste will be much less sensitive.
- Taste bud cells only last a week before the body renews them.
- You often lose your sense of taste when you get a cold because your nose is blocked and you can't smell—not because your taste buds have stopped working.
- Taste buds are not the only sense receptors in your mouth. There are also sensors for touch, pressure, moisture, heat, cold, and various other factors.

191

THE BODY'S DEFENSE SYSTEM

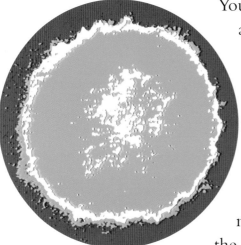

Your body is under frequent attack from bacteria, viruses, and many other microorganisms or "germs." To fight off such invasions, your body has developed a fantastic array of defense mechanisms, called the immune system.

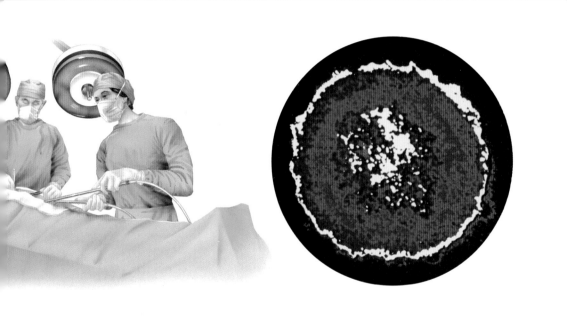

GERMS

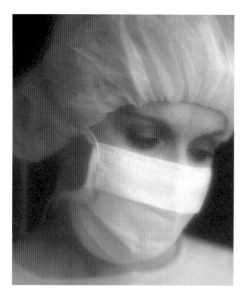

Most diseases are caused by tiny organisms, far too small to be seen except under a microscope. There are microbes like this living inside your body all the time. Most are completely harmless, such as the Escherichia Coli or E. Coli bacteria, which lives in your intestine. But some can damage your body by releasing toxins or interrupting the body's normal activities. Any organism that is harmful is called a pathogen—or, in common terms, a "germ."

When a colony of pathogens begins to multiply inside your body, you are suffering from an infectious disease. Infection normally sets off the body's immune system and many of the symptoms that you feel—fever, weakness, and aching joints—are often the side effects of your body's battle against the germ, rather than a direct effect of the germ itself.

CLEAN AIR
Microbes carried in the air can spread disease, so doctors working in hospital operating rooms wear face masks and cover their hair to cut the chances of passing on germs to the patient.

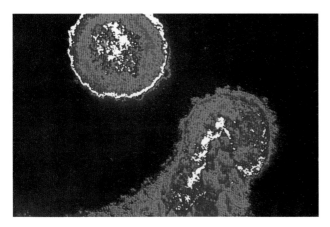

AIDS VIRUS
If you get Acquired Immune Deficiency Syndrome (AIDS), a virus called the HIV virus attacks the body's defense system. The virus gets inside crucial cells in the body's immune system, called T4 lymphocytes (see p200) and takes them over. As a result, the immune system is severely weakened when it comes to fighting off other infections.

KINDS OF INFECTION

Sometimes, an infection can spread throughout your body. This is called a systemic infection. Measles, colds, and the flu are all systemic infections. Sometimes, though, the infection is restricted to a small area. If dirt gets into a cut, for instance, this may infect the area around. This is called a localized infection.

Kinds of pathogen

Bacteria

- Bacteria are by far the most common form of microorganism. There are thousands of different kinds, but all are made from just one cell, and they can multiply rapidly.
- Pathogenic bacteria are grouped into three main shapes: round "cocci," rodlike "bacilli" and coil-like "spirilla."
- Diseases caused by bacilli include whooping cough, tetanus, typhoid, and TB.

Viruses

- Viruses are much smaller and simpler than bacteria. They live and multiply by taking over other cells, and cannot live outside.
- Viral diseases include colds, flu, mumps, rabies, and AIDS.

Parasites, protozoa, and fungi

- Parasites like tapeworms and amoebas can make you ill.
- Fungal spores and organisms called protozoa can also be pathogens.

LYMPHATIC SYSTEM

Without a good drainage system, your body could not stay healthy for long, and the lymphatic system is the body's drainage system. All body tissues are continually bathed in a watery fluid that comes from the blood. Much of the fluid drains straight back into the blood, but the rest, along with any other

discharge put out by the cells, such as bacteria and waste chemicals, drains toward the heart through the "lymphatics," the pipes of the lymphatic system. The lymphatics have no pump-like blood circulation, but rely on the continual movement of body muscles to push the fluid along. One-way valves ensure it only flows in one direction.

THE LYMPH SYSTEM

Here and there along the lymphatics are the system's waste treatment plants or "nodes." Lymph nodes are basically filters that trap germs and other foreign bodies that have gotten into the lymph fluid. To deal with the germs, the nodes have armies of "lymphocytes," a kind of white blood cell which can neutralize or destroy germs. When you have an infection, the lymph nodes may well swell as the lymphocytes battle with the invading bacteria or virus.

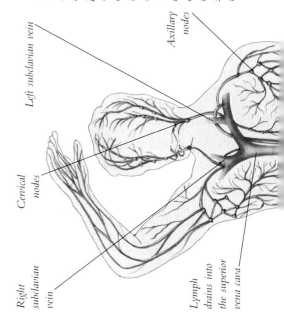

Left subclavian vein

Cervical nodes

Right subclavian vein

Axillary nodes

Lymph drains into the superior vena cava

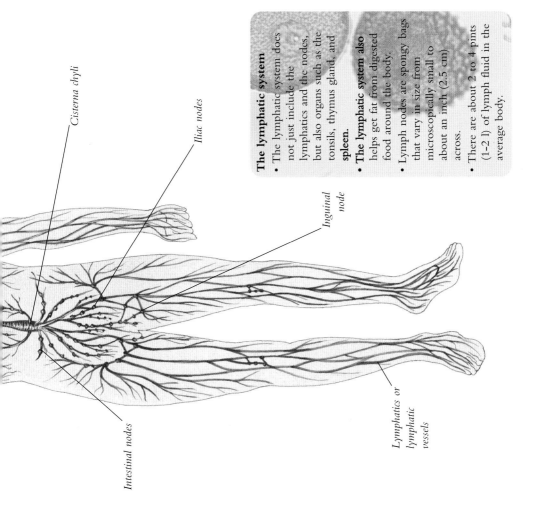

Cisterna chyli

Iliac nodes

Inguinal node

Intestinal nodes

Lymphatics or lymphatic vessels

The lymphatic system

- The lymphatic system does not just include the lymphatics and the nodes, but also organs such as the tonsils, thymus gland, and **spleen.**

- **The lymphatic system also** helps get fat from digested food around the body.

- Lymph nodes are spongy bags that vary in size from microscopically small to about an inch (2.5 cm) across.

- There are about 2 to 4 pints (1–2 l) of lymph fluid in the average body.

197

IMMUNE SYSTEM

Staying alive and well is a very difficult task for an organism as complex as a human being. So your body has developed an amazing array of systems for protecting itself against the microscopic invaders commonly called germs. First of all there is an array of barriers, toxic chemicals, and other booby traps to prevent the germs from getting into the body, including the skin and the mucus lining of the airway. Secondly, there are a host of special cells that actively attack the germs and try to neutralize or destroy them. Together, these two kinds of protection are called the body's immune system.

The mouth is protected by saliva, which contains the bacteria-killing enzyme lysozyme

The airways and lungs are protected by a coating of mucus, which traps germs, allowing cilia (tiny hairs) to sweep them out

The stomach is protected both by a coating of mucus and stomach acid

The eye is protected by tears, which wash away germs and contain an enzyme called lysozyme that kills bacteria

The inside of the nose is protected by hairs, which trap dust—and by sneezing

Inner defenses

- If germs get through the body's formidable physical and chemical barriers, the body has an arsenal of weapons to fight them off.
- In the blood, a mixture of liquid proteins called complement is activated by bacteria and attacks them.
- Proteins called interferon released by cells attack viruses, and also stimulate killer cells (see p200).
- Germs may be attacked by naturally "cytotoxic" (cell-poisoning) white blood cells.

Phagocytes

- Germs may be engulfed by big white blood cells called phagocytes—which literally means "eating cells."
- Phagocytes are attracted to wherever there is infection by chemicals released by inflammation (see glossary).
- When phagocytes encounter germs, they engulf them in a pouch in their membrane.
- Once engulfed, the germ is digested by enzymes.

The inside of the genitals is protected by mucus

OUTER DEFENSES

The body is equipped with a wide range of physical and chemical barriers to stop germs—harmful bacteria, viruses, and fungi—getting inside and spreading infection. Itching, sneezing, coughing, and vomiting are additional ways of getting rid of unwelcome material.

The sebaceous glands (see p64) ooze an oil that is toxic to many bacteria

As long as it is not broken, skin provides a good barrier against germs

FIGHTING CELLS

Many germs are actually quite similar to harmless bacteria—and even human body cells. So if your body's immune system was set up simply to wipe out all germs, it might actually end up damaging the body itself. So the immune system has a series of weapons that are designed to target particular invaders. Together these are called the adaptive immune system.

This is based on white blood cells called lymphocytes, which roam through the blood and the lymph system looking for trouble. There are two kinds—B lymphocytes and T lymphocytes. B lymphocytes or B cells use antibodies to target bacteria such as cholera. T lymphocytes or T cells are more important in the war against viruses.

1. Sooner or later, a germ bumps into its personal B cell—identified by matching proteins on its surface

B cells

Germ — bacteria

2. Once contact is made, the B cell multiplies, forming versions of itself called "plasma" cells

Plasma cells

3. Plasma cells make antibodies to attack the germ. The antibodies lock on to the target germ and make it easy for phagocytes to eat

Antibodies

4. Some B cells, called memory cells, go on multiplying after the germ has been wiped out—so if the germ returns, there are antibodies ready for it

THE B'S

B cells work using antibodies, tiny molecules that make germs easy prey for phagocytes, the eating cells that chomp up debris. There are thousands of different B cells circulating in the blood, each armed with antibodies against a particular germ. But there are only a handful of each until the one special enemy is contacted. If they do ever meet their enemy, they multiply dramatically at once and release hordes of the right antibodies.

Invaded cell

1. An invaded cell gives itself away with abnormal proteins on its surface

Helper

Killers

2. When helper cells meet abnormal proteins, they send out chemicals called lymphokines that tell killer cells to multiply

3. The killer cells lock on to the abnormal cell and destroy it

4. Some killer cells stay around, ready to attack any more abnormal cells

INVASION

Viruses are tricky germs to combat because they penetrate body cells and take them over. It is the T cell's job to root out and destroy their unfortunate victims.

THE T'S

Many germs, like viruses and parasites, hide inside, or take over your own body cells. It is the task of the T cells to identify and destroy these lost cells. There are two kinds of T cell: "killers" and "helpers." Helper cells are the ones that identify the invaded cells and send out the alarm. Killers are the ones that move in and destroy them.

Immunity

Targets

• B cells are effective against many bacteria such as cholera, some viruses such as measles, and a few parasites such as malaria.

• T cells are effective against many viruses such as 'flu viruses, some bacteria such as TB, and a few fungi such as candidiasis.

Immunity

• You only suffer from some infections once, because once is enough to give you "immunity."

• You are immune to a second infection because once activated, both B and T cells leave behind memory cells, versions of themselves ready to spring into action should the same germ return.

• Some immunity is "innate," which means you were born with it. Some immunity is "acquired"— either by exposure to the germ during an infection or by immunization (p202).

201

IMMUNITY

Remarkably few germs are likely to catch your body completely by surprise. Some germs your body is prepared to fight with its immune system even before you are born—with the right B cells and antibodies (see p201) sitting in wait for the enemy. Others you acquire immunity to during the course of your life.

One way you can acquire immunity is to catch the germ. Once your body has fought off the germ, it is properly prepared with a store of primed B cells and antibodies to fight it again.

Another way of becoming immune is by immunization (also called vaccination). Immunization means you are deliberately exposed to a mild or dead version of a germ—usually by a small injection—so your body builds up antibodies, and is resistant to further infection by the real germ.

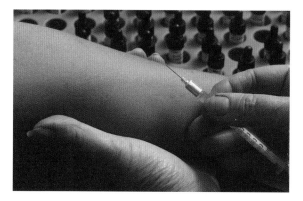

VACCINATION

Many once major diseases such as diphtheria, polio, measles, and whooping cough are now quite rare in many countries thanks to mass vaccination—the vaccination of a huge number of people, especially children. One once widespread disease, smallpox, has almost been wiped out. But there is a risk they might flare up again, so most children still go through a series of vaccinations.

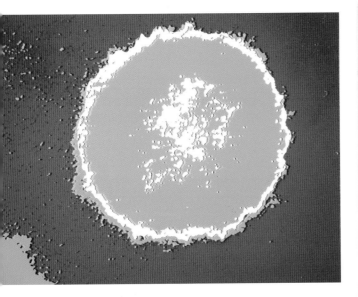

Immunization

Typical program for childhood vaccinations

- Three months: Diphtheria, whooping cough, and tetanus in a combined injection, plus polio on a sugar lump.
- Five months: Diphtheria, whooping cough, and tetanus in a combined injection, plus polio on a sugar lump.
- Nine months: Diphtheria, whooping cough, and tetanus in a combined injection, plus polio on a sugar lump.
- 15 months: Measles, mumps, and German measles in a combined injection.
- Four to five years: Diphtheria and tetanus in a combined injection, plus polio on a sugar lump.

Foreign things

- You may also have vaccinations if you are traveling abroad, against diseases such as yellow fever.

Passive immunization

- In passive immunization, antibodies from someone else are injected to provide immediate, short-lived protection.
- Your are injected with blood that is taken from someone who has already been exposed to the germ.

Active immunization

- In active immunization, your are given a killed version or otherwise harmless version of the germ.
- The germ provokes your body to make antibodies.
- When the real germ comes, then the body is ready.

203

REPAIRING THE BODY

Sometimes, when the body cannot heal itself and drug treatment is inappropriate, the only solution is to cut into the body and physically repair a problem. This is called "operative surgery." A minor operation is generally done under a "local anesthetic"—which means a drug is wiped on the skin or injected to kill any feeling in just the area being operated on. A major operation is done under a "general" anesthetic, which means the patient is given a drug by injection or breathes a gas that sends them completely to sleep for the entire operation. The biggest operations tend to be transplants of major organs such as the heart.

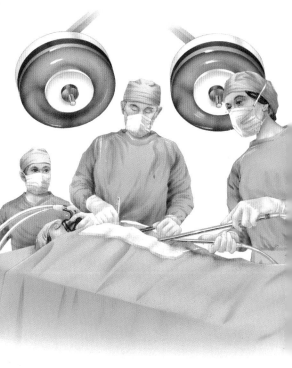

OPERATION
Major surgery involves a team of people working in a specially equipped room called an operating room. The team is headed by the surgeon, but also includes an anesthetist to ensure the patient stays unconscious.

204

HEART TRANSPLANT

If someone has very severe heart disease, they may have a heart transplant operation in which their diseased heart is swapped for the healthy heart of someone who has died. During the operation—which takes about 4 hours—the patient is connected to a machine called a heart-lung machine, which takes over the tasks of the heart while the heart is cut out. The new heart is stitched in, and after the patient comes to he is given drugs to stop his body's immune system from rejecting the new heart.

Meniscus _____

KNEE SURGERY

Surgery may be needed for all kinds of reasons. A football player, for instance, may have severely damaged the meniscus of cartilage in a knee joint by twisting his leg, and surgery may be the only way to repair it. With the patient under anesthetic, the surgeon inserts a viewing device called an arthroscope into the knee, drains out the synovial fluid (see p74) then takes out the damaged meniscus.

Surgery

Reasons for surgery include:

- Inspecting diseased or damaged organs or tissues.
- Replacing structures such as bones that have been damaged or grown wrongly.
- Redirecting blocked or diseased body channels such as heart arteries (heart bypass) or intestine (colostomy).
- Transplanting faulty organs with healthy organs from living or dead donors.

Other kinds of surgery

- In microsurgery, surgeons carry out tiny operations under a microscope—to rejoin blood vessels in the eye, for instance.
- In laser surgery—used for eye operations and stomach ulcers—the surgeon uses a laser beam, not a knife. The laser seals off blood vessels as it cuts.
- Endoscopic surgery involves the surgeon operating through an endoscope—a tube the surgeon puts into the body to see inside.

NEW BODY PARTS

Normally, given time, your body can heal most injuries and repair most damage done by disease. But occasionally, the damage is beyond even your body's remarkable capacity for self-repair. When this happens, different parts of the body may need to be replaced altogether.

Body parts can be replaced by transplanting. To repair a bad burn, for instance, skin may be taken from a healthy part of the body and grafted onto the injured place. Body parts, such as a heart or bone marrow can also be transplanted from another person, called a donor.

Sometimes, though, human body parts cannot be used, or are not available, and the replacement part is manufactured. This is called a prosthetic.

PROSTHETIC ARMS

Sometimes artificial limbs may have to be fitted on someone who has lost an arm or a leg through disease, injury or a birth defect. Artificial limbs are usually tailor-made for the patient, and a mold is made of the patient's stump so that the new limb can fit snugly. Often arms have hands with claws or fingers worked by electric motors. Electronic circuitry may be used to pick up muscle and nerve impulses in the stump. So the person only has to think of it for the electric hand to grip or move as they want, just like an ordinary hand.

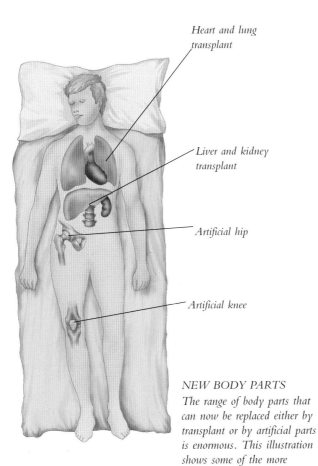

Heart and lung
transplant

Liver and kidney
transplant

Artificial hip

Artificial knee

NEW BODY PARTS
The range of body parts that
can now be replaced either by
transplant or by artificial parts
is enormous. This illustration
shows some of the more
common replacements.

Transplant surgery
Parts that may be replaced by
transplant
- Kidney
- Cornea of the eye
- Heart; heart and lung; lung
- Liver
- Pancreas
Rejection
- The main problem with
  transplants is that the body's
  immune system identifies
  the new organ as foreign—
  and so employs all its
  resources to attack it. This
  is called rejection.
- To reduce the chances of
  rejection, transplant patients
  are given the drug
  cyclosporin, which
  suppresses the immune
  system.
Storage
- After removal from the
  donor, the organ to be
  transplanted is washed in
  oxygenated fluid and
  cooled to reduce the
  possible damage due to lack
  of blood. It may be carried
  in a cool box.

New Life

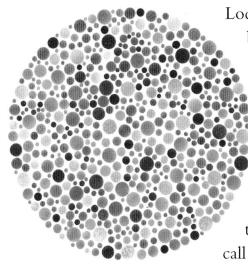

Locked within a few cells in your body, so small that they can only be seen under incredibly powerful microscopes, are the chemical instructions to make an entirely new human being. You were made when your mother and father bought their sets of chemical instructions together in the process scientists call "sexual reproduction."

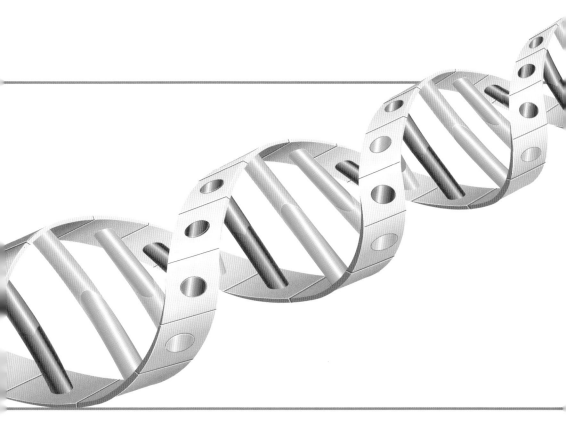

THE GENETIC CODE

Inside every cell in your body is a long thin molecule so small it can be seen only under the most powerful microscopes. This tiny molecule, called DNA or dioxyribonucleic acid, is actually all the instructions the cell ever needs to perform its tasks—and all the instructions needed to make a whole new human being just like you.

Like a computer, DNA carries all these instructions in a code, called the genetic code, and the code comes from its unique structure.

It is usually coiled up in a knot, but it is actually a very long molecule, made from two thin strands wrapped around each other in a long spiral called a double helix, a bit like a twisted rope ladder. The key to the code lies in the chemicals called bases that make up the rungs.

DNA MOLECULE

The DNA molecule has a unique double spiral shape, and carries complete instructions for life in coded form in the sequence of chemical bases that join its two strands together. Inside the cell, the molecule is coiled and twisted into threads called chromosomes, and it is these that hold your body's life instructions.

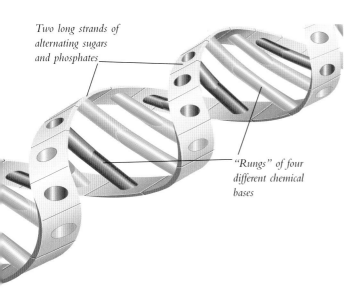

Two long strands of alternating sugars and phosphates

"Rungs" of four different chemical bases

THE FOUR BASES

There are four kinds of base in DNA—guanine, adenine, cytosine, and thymine. Each pairs only with one other—guanine with cytosine, adenine with thymine. So the sequence of bases along one strand of the DNA is a perfect mirror image of the sequence on the other. This means each can be used, like a mold, to make a copy of the other.

These bases are a bit like letters of the alphabet, and the sequence is broken up into sentences called genes, each providing the instructions to make a certain protein.

Genes

Building proteins

- Proteins are built from different combinations of chemicals called amino acids. These are all available in the cell, so to make a protein, the DNA simply has to instruct the cell as to which ones to stitch together.

Codons

- The bases in DNA are set in groups of three, called codons, and each spells out certain amino acids.
- There are 64 ways four bases can be ordered in groups of three.

RNA

- DNA is too valuable to use directly as a mold. So when the cell needs a new protein, it sends an enzyme to copy just the right bit on to a similar chemical called m(messenger) RNA.
- mRNA makes the new protein in a part of the cell called the ribosome (see p30) by linking amino acids.

NEW CELLS

New human body cells multiply by splitting in half again and again. This is how you grow as a baby, and this is how body cells that are worn out are replaced by new ones. When cells divide, each of the new cells must be identical, and carry the same set of instructions. So before each cell divides, it makes a copy of the DNA in its chromosomes (see p210) to share between the new cells in a process called replication. When DNA replicates, its two strands slowly

MITOSIS
Mitosis is the way cells divide to create new cells or replace those that have worn out. Each new cell gets an identical copy of the original cell's chromosomes.

1. Before cell division begins, each chromosome is copied, and the two copies coil up into dark rods that join to make an X-shaped pair

2. The pairs of chromosomes line up across the center of the cell. They stick onto tiny threads that grow across the cell

3. The threads begin tugging in opposite directions. The pairs split in half, and each half is pulled to the opposite end of the cell

4. A new nucleus starts to form around the cluster of chromosomes at each end of the thread. Each cluster of chromosomes is identical.

5. A membrane grows around the two new nuclei, and eventually the old cell divides down the middle to create two brand new cells.

unzip so that mirror image sequences of bases are exposed on each strand. Then bases floating around in the cell called free bases latch on to the strands. Each of the four bases latches on to its natural partner—guanine with cytosine and adenine with thymine. In this way, a matching strand of DNA is created on top of each of the two strands from the original DNA. So there are now two identical copies of the DNA, and cell division can begin.

Germ cells

- You have two sets of chromosomes in your body cells—one set from your mother, the other from your father.
- So before you were conceived, both your father's and mother's bodies had to make "germ" cells with just one set of chromosomes.
- Germ cells are created in a unique way, different from mitosis. This is called meiosos, in which cells divide to give half the normal number of chromosomes.
- When you were conceived, a cell created by meiosos in your father joined with a cell created by meiosos in your mother to create a cell called a "zygote." This is how you began.

Meiosis

- Before meiosos begins, chromosomes in each pair swap genes (see p211). They then line up as in mitosis, and the cell divides to leave two cells with the normal number of chromosomes.
- These two new cells then split again as in mitosi—but because there has been no replication, each of the four new germ cells has just one set of chromosomes.

HEREDITY

You probably resemble your parents a bit. The genetic blueprint passed on by parents to children is so reliable that they are surprised if you don't. Yet no two people in the world are exactly alike— not even twins. This amazing balance between precision and waywardness in passing on characteristics comes from your chromosomes. There are 46 chromosomes in every cell (except germ cells, p213), and these can be matched into pairs. There are 23 matching pairs in women's cells, but just 22 in men's— two are odd. The odd pair is the sex chromosomes, called X and Y because of their shape. Women have these sex chromosomes, too, but have two matching X's, not an X and a Y. It is these chromosomes that say whether a child will be a boy or a girl. Two X's make a girl; an X and a Y a boy.

GIRLS WILL BE GIRLS
It was when your parent's germ cells came together that it was decided whether you would be a boy or a girl. Because women's cells have only X chromosomes, all their germ cells have X chromosomes too. But men's cells have an X and a Y. So when their germ cells form (see p213), half have an X and half a Y. So during fertilization, a woman's germ cell may join with a man's germ cell which contains either an X or a Y. If it is an X, the baby will be a girl; if it is a Y, the baby will be a boy.

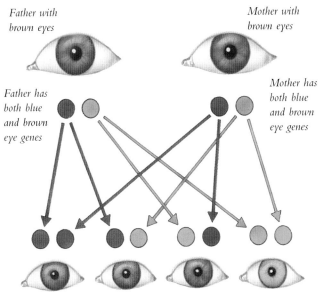

Father with brown eyes

Mother with brown eyes

Father has both blue and brown eye genes

Mother has both blue and brown eye genes

Two brown eye genes give brown eyes

One brown gene and one blue gives brown eyes

One blue gene and one brown gives brown eyes

Two blue eye genes give blue eyes

BLUE OR BROWN?

Brown eyes are often coded for by a gene that is dominant over a gene for blue eyes (see panel). Two parents with brown eyes, though, may have children with blue eyes if they carry the gene for blue eyes—as long as the child gets the recessive gene from both parents. Three out of four times, the child will get the dominant brown gene from one or both parents and so have brown eyes. But one in four times, the child will get the two recessive genes and have blue eyes.

How genes are passed on

- Apart from the sex chromosomes, all the other 44 come in pairs.

- In every pair, both chromosomes give instructions for the same thing, the same features are coded for by genes in the same place on each chromosome, called the gene locus. So you have two alternative instructions for each feature.

- Some chromosomes may carry many genes, called polygenes, to code for a single feature.

- Most features are a mix of the two sets of instructions. But with some, there can be no alternative—one or the other must win. The gene that wins is said to be "dominant;" the loser is "recessive."

- A recessive gene does not always lose. When the recessive gene is in each of your parent's germ cells, the recessive gene will have no competition.

MISPRINTS

The genetic code is remarkably reliable, passing on all the instructions for a human being from one generation to the next faultlessly. But every now and then, it makes a slip. Sometimes, this slip or "mutation," involves a single base in millions. At other times, it may be an entire chromosome that goes wrong.

Genes mutate when DNA is copied wrongly before a cell divides (see p212). This happens naturally, but the chances of a mutation are dramatically increased by exposure to radiation. The mutation may effect not only the growing child. If the child with the mutant gene lives to have children, the gene may be passed on down through the generations. Some mutations occur on the sex chromosomes—so boys in a family may get the gene and girls not, or vice versa.

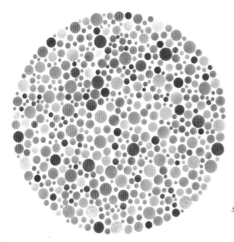

SEEING STARS

If you have normal color vision, you should see a green star in among the red dots. If you suffer from red-green color blindness, you won't. Red-green color blindness is a mutant gene that makes people unable to tell red from green. It is a recessive gene (see p215) carried on the X sex chromosome. The diagram on the right shows how a boy can inherit this gene and be color blind while his sister isn't.

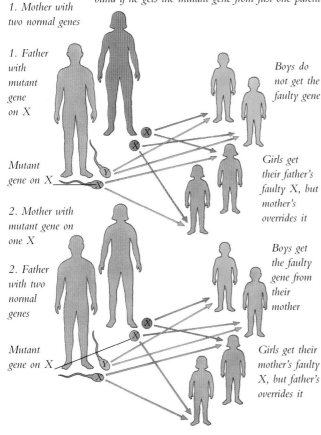

WHO WILL BE COLOR BLIND?

The red-green color blindness gene on the X chromosome is recessive, so a girl (who has two X chromosomes) will only be color-blind in the very rare event that she gets the mutant gene on the X chromosome from both parents. But a boy (who has only one X chromosome) will be color-blind if he gets the mutant gene from just one parent.

1. Mother with two normal genes

1. Father with mutant gene on X

Boys do not get the faulty gene

Mutant gene on X

Girls get their father's faulty X, but mother's overrides it

2. Mother with mutant gene on one X

2. Father with two normal genes

Boys get the faulty gene from their mother

Mutant gene on X

Girls get their mother's faulty X, but father's overrides it

Sickle-cell anemia

- Sickle-cell anemia is an inherited disease which is very common in Africa.
- Hemoglobin is the chemical that binds oxygen into red blood cells. In people with sickle-cell anemia, a mutant gene has given them the wrong kind of hemoglobin.
- Sickle-cell hemoglobin crystallizes whenever there is a shortage of oxygen in the blood. As it does, it pulls the red blood cells into a sickle-shape.
- Sickle-cells break up quickly, so a person with sickle-cells quickly becomes anemic as he loses red blood cells.
- The cause of sickle-cell anemia is just one faulty base in hundreds in the gene for hemoglobin.
- Sickle-cell anemia is common in Africa in the same area as the disease malaria is. The mutation that causes sickle-cells seems to protect people from malaria—so they live to pass on their faulty gene.

217

GROWING UP

From the moment you are conceived until you are in your late teens, your body continues to grow bigger and bigger. You grow fastest in your first few years, then grow slower, then go through a growth spurt in your early teenage years.

At some stages, girls grow faster, at others it is boys. Baby boys grow faster than baby girls—but only for the first seven months, then girls begin to shoot ahead until about the age of four. After that girls and boys grow at the same rate until puberty. Girls reach puberty first so grow taller than boys for a while. Finally, boys reach puberty and grow taller.

CRAWLING
The age at which children learn to walk varies hugely, but it is over nine months before most babies are strong enough to support their own weight. Once this happens, they learn to walk automatically, helped by a primitive reflex (see p138). This reflex makes the baby's legs move in a walking motion when it is held upright with the soles of its feet pressing on the floor.

BABY SKELETON
A baby's skeleton is actually quite soft. It has to be to allow room for growth. It is only as you get older that it gets longer and hardens and takes on its adult shape.

BIG HEADED BABIES

When you grow, you don't simply get taller and heavier. Different parts of your body grow at different rates, so you actually change shape. When you are born, your head is already three-quarters the size it is going to be when you're adult, and by the time you're one, it is almost full adult size. So your head appears to shrink as you grow older, while legs grow.

A baby's head may be a quarter of its total height

At eight, your head is about one-fifth your height

At 12, your head is about one-sixth your height

At 15, your head is about one-eighth your height

Growth

Growth rates

- Different tissues grow at different rates.
- Lymph tissues grows rapidly during early childhood.
- Your brain grows most quickly up until the age of five, by which time it is 90 percent of its adult size.

Growth hormone

- Your body is encouraged to grow in childhood by a hormone secreted by the pituitary gland (see p166) called growth hormone.
- Growth hormone stimulates cells to make protein and breaks down fat to make energy.
- Too much growth hormone can make you unusually tall or unusually short.
- Thyroid hormones and sex hormones also stimulate growth.

Growth rates

- If your parents are tall, you are likely to be tall.
- If your diet is poor, you are likely to be short.

PUBERTY

You are born with reproductive or sexual organs, but they only develop in the right way for you to have children once you reach the age of puberty—typically when you are 11 to 13 years old. At puberty, hormones sent out by the pituitary gland (see p166) flood through your body, stimulating the changes that turn boys into men and girls into women.

When a girl reaches puberty, she begins to grow breasts, and soft, downy hair sprouts under her arms and around her genitals. Soon her hips begin to grow wider and her waist slimmer. Eventually, she begins her monthly periods, or menstruation.

When a boy reaches puberty, his testes grow and change shape, and begin to produce sperm. At the same time, he begins to grow pubic hair and hair on his chin.

TEENAGE CHANGES
Puberty arrives at different ages in different people but girls typically reach puberty at 11 or so, two or three years earlier than boys, though it can often be as early as 9 or as late as 15.

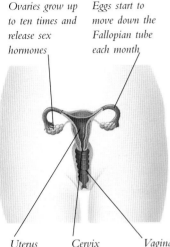

Ovaries grow up to ten times and release sex hormones

Eggs start to move down the Fallopian tube each month

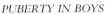

Uterus enlarges

Cervix

Vagina

PUBERTY IN GIRLS

When she reaches puberty, a girl's ovaries begin to grow invisibly inside her, up to ten times as big. As the ovaries grow, they release the hormones oestrogen and progesterone. Then eventually, around the age of 13, she experiences her menarche, her first menstrual period. Periods are erratic at first, but usually settle down to a regular 28-day cycle. When this happens, she is sexually mature and physically able to have a baby.

PUBERTY IN BOYS

When a boy reaches puberty, his sex organs begin to develop inside. The seminiferous tubules in his testicles begin to make sperm. The testicles hang farther outside the body where it is cool enough for sperm to thrive. By the time he is 15 or so, his body can create 200 million new sperm a day.

Vas deferens

Urethra

Bladder

Penis grows

Scrotum drops

Testicles begin to make sperm

Chemical triggers

- The changes that come with puberty are triggered by rising levels in the blood of two hormones: follicle stimulating hormone (FSH) and luteinizing hormone (LH).
- FSH and LH both come from the pituitary gland.
- The command to the pituitary to send out FSH and LH comes from the hypothalamus (see p166).
- It may be that the command goes out from the hypothalamus when it has grown the right nerves.
- When levels of FSH and LH in the blood are high enough, a girl's ovaries and a boy's testicles grow rapidly.
- As a girl's ovaries grow, they send out the hormones oestrogen and progesterone, which produce the bodily changes of puberty.
- As a boy's testicles grow, they send out the hormone testosterone, which produces the bodily changes of puberty.

MONTHLY RHYTHMS

Every 28 days or so, a woman's body goes through a regular cycle of changes called the menstrual cycle. They affect the whole body, but their purpose is to prepare one of the woman's eggs or ova for fertilization.

Only a few eggs are ever fertilized, so the cycle normally ends with the shedding of the egg and other

preparations, ready for a fresh start next month.

Unlike male sperm, a woman's sex cells (called oocytes) are ready formed at birth. Each is stored in a minute sac called a follicle in the woman's ovary. The cycle begins when Follicle Stimulating Hormone (FSH) sent by the brain's pituitary gland prods several follicles to start growing.

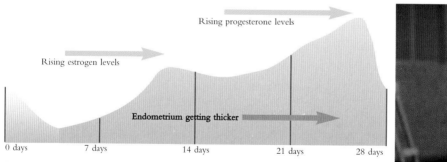

Rising progesterone levels

Rising estrogen levels

Endometrium getting thicker

0 days   7 days   14 days   21 days   28 days

THE MENSTRUAL CYCLE
During the 28-day menstrual cycle, as an egg is prepared inside the ovary, the "endometrium" lining of the uterus gradually gets thicker, ready to receive a fertilized egg in response to the hormones estrogen and progesterone, as this graph shows.

THE SEX HORMONES

As the follicles grow, they release another hormone—the sex hormone estrogen—which spurs the lining of the uterus to thicken into a rich bed for a fertilized egg, called the endometrium. When an egg is ripe, it bursts from its sac, and slides down the Fallopian tube toward the uterus. This is called ovulation.

If the egg is fertilized by sperm in the Fallopian tube, the uterus lining goes on thickening ready for the pregnancy. If it is not, the egg, along with the endometrium, is shed—so that blood and fragments flood out from the woman's vagina in what is called menstrual flow.

Hormone release
Four key hormones:
- FSH from the pituitary starts follicles ripening.
- Estrogen from the follicle thickens the uterus lining and prompts the release of luteinizing hormone (LH) from the pituitary.
- LH spurs a ripe egg to burst out of its follicle and into the Fallopian tube.
- Progesterone from the corpus luteum (see below) carries on thickening the uterus lining.

Corpus luteum
- After it has lost its egg, the abandoned follicle turns from white to yellow and is called a corpus luteum (which means "yellow body").
- The corpus luteum sends out progesterone.
- If the egg is not fertilized—about 22 days into the cycle—the corpus luteum begins to waste away and levels of the sex hormones estrogen and progesterone drop, ending the cycle.

NEW LIFE

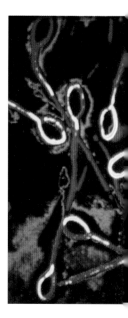

For a new human life to be created, a sperm from a man must fertilize an ovum or egg from a woman. Only when the 23 chromosomes in each are united to give the full complement of 46 will a baby begin to develop. Fertilization can now actually take place in a laboratory dish, but normally it occurs during sexual intercourse between a man and a woman.

At the climax of intercourse, millions of tiny sperm are ejaculated from the man's penis into the woman's vagina. The sperms then swim up the uterus towards the woman's Fallopian tube (see p222), where there may be an egg waiting. Very few sperm make it to the right tube, but only one is needed to fertilize the egg.

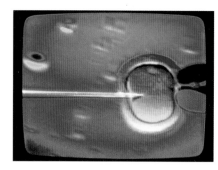

TEST TUBE FERTILIZATION
To fertilize an egg in a laboratory, an egg is taken from the potential mother's ovaries and joined with a sperm from the father. Once the egg is fertilized and starts to divide, it is usually returned to the mother's womb. This is called "in vitro" fertilization because it happens in a glass dish—"vitro" is the Latin for glass. In vitro fertilization can often lead to multiple births—twins, triplets, and even sextuplets.

EGG AND SPERM
Of the hundreds of
millions of sperm that are
ejaculated during intercourse,
only one needs to reach the egg to
fertilize it. The sperm is much smaller than the
egg, and fertilizes it by burying its head in the
egg's huge side. As it is engulfed by the egg, its
tail drops off and other sperm are shut out. Soon
after, the egg divides, and prepares to unite with
the sperm to form the first cells of the new life.

Sperm and egg race

• Sexual intercourse does not always result in a new life being conceived. The egg can only be fertilized in the 12 to 24 hours immediately after the woman ovulates (see p222).

• For the egg to be fertilized, a man's sperm must reach the woman's egg when it is moving down the Fallopian tube, halfway through the woman's menstrual cycle.

• When sperm are ejaculated during intercourse, they land near the woman's cervix, the neck of her uterus.

• Sperm look like tadpoles, and they swim up the uterus by waggling their tails.

• As the sperm approach the egg, their heads dissolve in chemicals released by the egg. But as they do, they send out an enzyme that helps one penetrate the egg. It takes the enzymes from hundreds of sperm to break down the egg's protective barrier—but one may finally make it through.

PREGNANCY

Pregnancy begins the moment an ovum or egg is fertilized by a sperm. The egg then divides inside the mother to make more cells and grows into an embryo, and after eight weeks, into a fetus. Unlike an embryo, a fetus has limbs and internal organs and it grows rapidly, protected inside its mother's uterus, now called the womb—and the womb grows with the fetus. After around nine months, the fetus is fully developed and ready to be born.

2. 5 weeks
Arm and leg buds appear

1. 4 weeks
Heart starts to beat

4. 12 weeks
The head is now quite large, though the eyes are closed. Finger and toe nails start to grow

6. 20 weeks
If the fetus is a boy, its genitals may be visible. Lungs are formed, and the fetus can grip

3. 8 weeks
Embryo becomes fetus as limbs form. Fingers and toes are webbed

5. 16 weeks
The fetus is now recognizably a baby, covered in downy hair called lanugo. It may begin to kick

226

Womb

Fetus

Cervix

Amniotic sac

Placenta

Umbilical cord

IN THE WOMB

Inside the womb, the new human life grows from just a few cells to an embryo, then a fetus to become a baby in nine months, growing larger and more recognizably human by the day. Through this time it is kept warm and cushioned from the outside world inside a bag or sac of fluid, called amniotic fluid. It is supplied with oxygen and food in blood via the umbilical cord. This links the fetus to the placenta, a spongy organ lining the womb.

SOUND SCAN

16 to 18 weeks into pregnancy, an expectant mother may go for an ultrasound scan. This uses sound waves too high for humans to hear to create a picture of the inside of the womb. The fetus can be seen moving on a display screen if it is healthy.

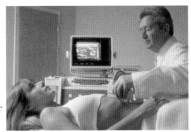

Pregnancy

- A woman can often tell that she is pregnant because her monthly period (see p222) fails. A urine test confirms the pregnancy.
- Pregnancy is divided into three periods or "trimesters:" 0-12 weeks, 13-28 weeks, and 29-40 weeks.
- In the first 6–8 weeks, a pregnant woman is often sick in the morning.

Body changes

- During pregnancy, a mother's body changes shape to meet its demands.
- She gains 30 percent more blood.
- Her heart rate goes up.
- She eats more and puts on weight. Surprisingly, less than half this extra weight is the fetus itself.
- Her belly begins to expand.
- The hormone estrogen makes the breasts grow larger and develop milk glands.
- The hormone progesterone relaxes the muscles of the abdomen to allow the womb to expand.

227

BEING BORN

By the end of nine months, a fetus is fully developed. It turns upside down with its head pointing into the entrance of the womb, ready to make an entrance into the world. Birth begins when the muscles of the uterus begin to contract rhythmically to push the baby out through the cervix. This muscle contraction is called labor. The process for the mother is often long and very painful, but eventually the baby is pushed forcefully into the world. It emerges through the cervix and vagina which expand to make what is called the birth canal. When the baby finally makes it out of the canal, the shock of the new world normally starts the baby breathing through its own lungs for the first time. The umbilical cord linking it to its mother can then be cut.

MOTHER AND BABY
Although physically separated from its mother, a newborn baby still relies on her entirely. For a few months at least, a newborn baby sucks milk from her breasts. All the baby's early food may come from breastmilk, as well as a number of vital protective chemicals, and breastfeeding plays an important part in establishing the relationship between mother and child.

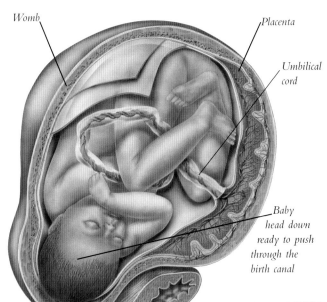

Womb

Placenta

Umbilical cord

Baby head down ready to push through the birth canal

Cervix

Birth canal

CHILDBIRTH

When the baby is about to be born, the amniotic sac bursts, releasing its fluid. When the cervix widens to about 4 inches (10cm) across, the birth is about to happen. The womb then contracts, pushing the baby through the birth canal headfirst out into the world. It is still attached to the placenta by its umbilical cord. The doctor usually cuts the cord as soon as the baby emerges.

Childbirth
Stages of labor

• There are three stages in childbirth, usually called the three stages of labor.

• In the first, the muscles of the womb begin to contract and the birth canal widens. The contractions eventually burst the amniotic sac, letting its fluid run out through the vagina. This is called the breaking of the waters.

• The second stage is the birth itself, when the baby passes out through the birth canal and the umbilical cord is cut.

• In the third stage, the placenta is shed and follows the baby out through the birth canal. This is called the afterbirth.

Premature babies

• Sometimes babies are born "premature"—before they are fully developed—and may have to be sustained in an incubator until they are big and well enough to cope by themselves.

229

OLD AGE

By the time you reach the age of 20 or so, you are fully grown. From that time on, your body starts to deteriorate, slowly at first but increasingly rapidly in later years.

People age at different rates, but by the time you are 65 or so, you will certainly show all the signs of a body in old age. Your hair will probably be gray or white. You may go bald if you're a man. Your skin will be wrinkled. You may stoop. And you will probably find all your senses less sharp.

No one knows quite why we do get older. There are many ideas. It may be that every cell has a built-in lifespan program. Or it may be that the body's genetic coding system is like a record played again and again. The more it is played, the more "scratched" and unreliable it gets.

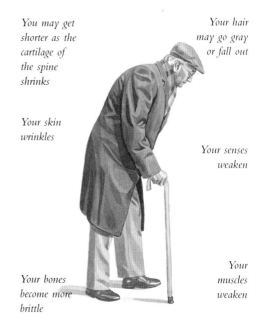

You may get shorter as the cartilage of the spine shrinks

Your skin wrinkles

Your hair may go gray or fall out

Your senses weaken

Your bones become more brittle

Your muscles weaken

SIGNS OF AGING

As you grow older, your body will stop working quite as well as it did when you were young. Your immune system becomes weakened, too, and you are more likely to succumb to diseases you may have fought off when younger.

WEARING OUT?

Many more people live to an old age nowadays, and remain fit and active, thanks to better health care and nutrition. Most men can expect to live until they are at least 70, and women until they are 76. Yet the oldest people are no older now than they have ever been.

HARDENING ARTERIES

Aging tends to weaken the circulation and breathing. Blood vessels tend to become stiff and restricted by fatty deposits, a process called atherosclerosis. To force blood through the arteries, the heart must pump harder, raising blood pressure. A man of 70 may have 40 percent higher blood pressure than a man of 25. This puts a strain on the heart and cuts blood flow to tissues—and makes some old people forgetful and easily upset if the blood supply to the brain is affected.

Aging

Hair and skin

- Hair goes gray as pigment cells stop working.
- Skin becomes wrinkled as collagen fibers that support the skin stretch and go slack.
- Ultraviolet light from strong sunlight accelerates collagen slackening. So wrinkles appear first in the areas of skin exposed to the sun—like the face and hands.

Muscles and bones

- Muscles weaken as fibers die but are not replaced.
- Bones lose calcium and become more brittle and liable to break.
- As cartilage between their bones begins to dry out and shrink, old people's joints become stiff and they may begin to stoop.

Senses

- Old people's eyes take longer to adapt to changes and become poorer at focusing.
- Old people cannot hear so well, especially high sounds.
- Old people start to appreciate only stronger tastes.

231

HUMAN
DEVELOPMENT

Human beings have not
always been around. Indeed
compared with many
animals, humans have only
been around for a short
time. Fossils show that the
first humanlike creatures
appeared on Earth under 6
million years ago, and
humans like us barely
30,000 years ago.

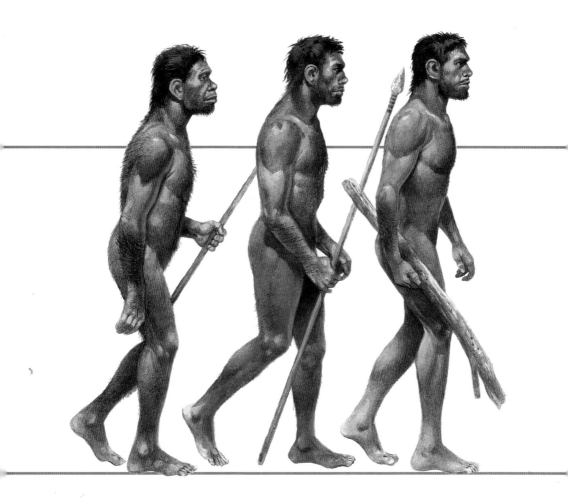

OUR ANCESTORS

We humans have a great deal in common with apes—including their long arms and fingers and big brain. So scientists believe we share the same ancestor. This ancestor probably resembled an orangutan and lived in the grasslands of Africa.

Just when our ancestors became different from apes is not known, for there are few remains to provide any clues. But humans are so similar biochemically to gorillas and chimpanzees that most scientists think it cannot be more than six million years ago.

The oldest remains of hominids (humanlike creatures) date from around 3.5 million years ago. All these early remains are called Australopithecus, which means "southern ape," because the first of them was discovered in South Africa.

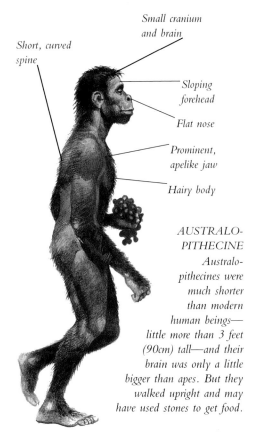

Short, curved spine

Small cranium and brain

Sloping forehead

Flat nose

Prominent, apelike jaw

Hairy body

*AUSTRALO-PITHECINE
Australo-pithecines were much shorter than modern human beings—little more than 3 feet (90cm) tall—and their brain was only a little bigger than apes. But they walked upright and may have used stones to get food.*

LUCY

"Lucy" is the most famous of early hominid remains. Lucy was a female Australopithecine who lived in Africa around 3 million years ago. Remarkably, her skeleton was almost complete. It showed conclusive evidence that Lucy was "bipedal"—that is, she walked on two legs, the first step on the road to being truly human. She was named Lucy after the old pop group the Beatles' song "Lucy in the Sky with Diamonds," which was playing on the radio at the time she was discovered in 1974.

MARY LEAKEY

Lucy was discovered by Don Johanson and Maurice Taieb, but many of the great discoveries of early hominid remains in Africa have been made by members of the Leakey family, including Louis Leakey, his wife Mary, and their son Richard. It was Louis Leakey who discovered Handy Man in 1931 (see p236).

The earliest hominids

- Lucy belongs to the group *Australopithecus afarensis*. But there were other groups of Australopithecine, such as *Australopithecus anamensis*, dating from 4.1 mya (million years ago).
- In 1995, Tim White found remains of a very old hominid, called *Ardipithecus ramidus* in Ethiopia—dating from 4.4 mya.
- In 1999, even older hominid remains were found in South Africa, but the "missing link" between apes and humans has yet to be found.
- Australopithecus and most of the earliest hominid remains are not our direct ancestors. They belong to branches of the hominid family that died out.

Different bodies

- Early hominids' heads were a different shape from ours. they had a prominent jaw, a large mouth, a low forehead, and a flat nose, like an ape's.

HANDY MAN

It was about two million years ago that the first really humanlike creatures appeared. They were much taller than Australo-pithecines, and had noticeably bigger brains. They used stone tools to cut hides for clothes and meat for eating, and also built simple shelters. The best known example is one called *Homo habilis*, which means "handy man," remains of whom were found at Olduvai Gorge in Tanzania, but there may have been others.

Bigger cranium

Flatter face

Straighter forehead

Straighter back

Good grip

GETTING TALLER
Australopithecines were little taller than an 11 year-old today, but Handy Man was notably taller — about the height of today's 13 year-old. He also walked considerably more upright than his ancestor.

236

Australo-
pithecine

Homo
habilis

HANDY MAN'S HOME

Remains of Handy Man were found in Olduvai Gorge in Tanzania, Africa, one of the richest sites for early hominid remains. It was here that Louis Leakey found some hand bones, a jaw, and skull fragments in 1961. From these, he was able to piece together a picture of Handy Man.

STONE TOOLS

Some scientists believe the first hominid tools were made of bone. But all early hominids used stone and wood, which is why the earliest period of human prehistory is usually called the Old Stone Age or Paleolithic ("paleo" means old and "lithic" means stone).

HANDY HAND

One of the things that Handy Man had that his ancestor didn't was the hand that earned him his name. It combined a power grip like that needed to wield a hammer, and a precision grip like that needed to write with a pen. This precision grip depended on having a thumb that could be rotated to meet the tip of another finger—called an "opposable thumb"—as we can with ours.

Skull facts

- Skulls play an important part in working out how human-like creatures have evolved, gradually becoming less ape-like and more human.
- Handy Man was the first really big-brained hominid. Australopithecus had a brain of about one third the size of yours. Handy Man's was about two-thirds the size of ours.

Could they speak?

- Scientists are divided about which early hominids, if any, could speak.
- Evidence for speaking comes from the shape of the skull and throat.
- A lump in the skull that may indicate a Broca's area (see p152) may show that Handy Man could speak.
- To speak, early hominids would had to have developed the same wide pharynx (throat, p43) that we have. This only came with walking upright—so Australopithecines probably could not speak.

WALKING UPRIGHT

A little under two million years ago, a third major kind of hominid appeared. This one, called *Homo erectus*, walked upright and was as tall, if not taller than most modern humans. *Homo erectus* could light fires, cook food, and hunt with wooden spears. For the first time, hominids moved out of Africa and spread as far as Russia and Indonesia, which is why he is sometimes known as Java man. Australopithecus was more like an ape, but Homo erectus was very humanlike in every way.

Flatter forehead as brain expands

Jaw becomes smaller and flatter

Taller, upright posture

Narrower waist as guts shrink

GETTING TALLER

Australopithecines were very small and Handy Man was no bigger than a teenager. But Homo erectus was as tall, if not taller, than most people today. He walked completely upright, which is how he got his name, Homo erectus. However, his brain was much the same size as Handy Man's, so he was not very intelligent.

Australo-
pithecus

Homo
habilis

Homo
erectus

238

FIRE-RAISING

Learning to light fires was one of Homo erectus's great achievements. It not only kept him warm and allowed him to live through winter in much colder places. It also meant he could cook food. Cooking softens and makes edible food that simply cannot be eaten raw. So hominids could make use of a much wider range of foods—and they did not need such large guts to digest the raw plant fibers that Australopithecines managed on.

THE FIRST HUNTERS

Australopithecines were probably just vegetarians like apes, but Homo erectus was a hunter, and this may have been crucial to the growth of the brain. Big brains need plenty of high-quality protein, and meat provides this. Moreover, to hunt effectively, hominids must have had to work together, and this interaction may have been one of the spurs that helped early hominids to develop language.

Homo erectus

- Remains of *Homo erectus* have been found all over Africa and Asia.
- The oldest remains are those from Swartkrans in South Africa and from Sangiran in Java, which date back 1.8 million years.
- Although short-lived *Homo habilis* died out, *Homo erectus* probably lived alongside Australopithecines for a million years or more.

Tools

- *Homo habilis* used sharp-edged pebbles as tools, but *Homo erectus* learned how to chip flakes off flints with another stone to give a razor-sharp edge.
- Flint hand-axes were probably used for butchering large animals caught in traps.
- Later, *Homo erectus* learned to fashion blades from flint by hammering on a bone or wood punch with a stone. These blades could be tied to a wooden shaft to make a spear.

239

WISE HUMANS

Some scientists believe that the similarity of the DNA in our bodies (see p210) shows that all humans today may be descended from a single woman, nicknamed "Eve," who lived in Africa about 200,000 years ago. Others say that because *Homo erectus* remains are found all over Eurasia, humans must have evolved separately in different places.

We belong to a group of species called *Homo sapiens* or "Wise Man," who first appeared about 100,000 years ago. The best known of early Wise Humans was *Homo sapiens neanderthalis* or Neanderthal man. He had a large face and rugged body, had a bigger brain than ours and wore clothes.

HUMANS ARRIVE

Humans are little, if any, taller than their ancestors Homo erectus and Neanderthal. In fact, most humans are actually shorter than any Homo erectus who lived in Africa. In hot climates, it pays to be tall and thin so you lose heat easily. Many humans are shorter partly because being short helps you cut heat loss. Poor nutrition also makes people short.

240

Neanderthal man

Australo-pithecus

Homo habilis

Homo erectus

Neanderthal man

Human

CAVE PAINTING OF A BISON
Both Neanderthals and early humans left
behind beautiful paintings on cave walls.

MODERN HUMANS

The first humans like us, called
Homo sapiens sapiens, appeared
just 30,000 years ago. The first
humans lived alongside
Neanderthal man for some
time—and may even have
lived alongside *Homo erectus*,
just as *Homo erectus* lived
alongside Australopithecus.
No one knows quite why
the others died out.

Cro-Magnon
man

Homo sapiens
Neanderthals
- There were Ice Ages in the
  time Neanderthals lived in
  Eurasia, but they coped
  with the cold by living in
  caves and skin tents and
  weaing clothes.
- They had well-made flint
  tools, knives, and scrapers.
- They buried their dead.
- They had much broader
  heads, with sloping
  foreheads and more
  prominent jaws. They may
  have had hairier bodies.

Homo Sapiens
- The first modern humans
  are called Cro-Magnon
  man, from a cave in France
  where remains were found.
- Cro-Magnon man probably
  emerged first in Asia, then
  spread across the world.
  About 12,000 years ago
  they lived in Australia and
  the Americas.
  - Cro-Magnon man
    looked like us, with a
    much flatter face than his
    ancestors, a prominent nose
    and small jaw.

EXERCISE

To stay in shape, your body needs regular exercise. The following pages show you just what exercise does.

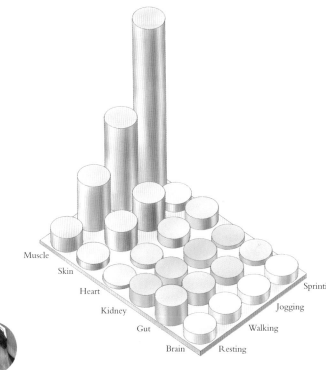

Muscle
Skin
Heart
Kidney
Gut
Brain
Resting
Walking
Jogging
Sprinting

BLOOD DIVERSION
The body adapts to exercise by diverting blood to the muscles. If you go for just a gentle stroll, this will have little effect on the rest of the body. But if you exercise hard, the blood supply to the muscles goes up sharply while the supply to your heart drops. Only the supply to the brain stays steady. The height of the columns (left) shows how much blood goes where with different kinds of exercise.

STAYING IN SHAPE
Exercise only has real benefits if it lasts long enough to be "aerobic." This means that the muscles have got to the stage where they are using oxygen to burn glucose, not just burning glucose as they do when you start exercising. Regular aerobic exercise improves fitness by building up the body's capacity to supply the muscles with oxygen by strengthening the heart.

Immediate effects of exercise
Your body responds to strenuous exercise in these ways:

- The heart rate doubles and so does the volume of blood that it pumps. This boosts heart output by five times or more.
- You breathe faster and deeper, so the amount of air you take in every minute goes up by ten times.
- Blood vessels in active muscles get four times wider, while blood vessels in the gut and skin contract, diverting blood to the muscle.
- The rate at which energy is used in the muscles goes up by about 20 times.
- Insulin levels drop and adrenalin rises, so the liver converts glycogen to glucose.
- Body temperature rises and you begin to sweat. The blood vessels in the skin which at first constricted begin to dilate.
- Lactic acid builds up in the muscles as they burn glucose without oxygen (see p93).

Recovery

- If you're in shape, all systems return to normal within minutes after light exercise.
- After heavy exercise, it can take hours for your body to return to normal—mainly because your body has to oxidise all the lactic acid that has built up.

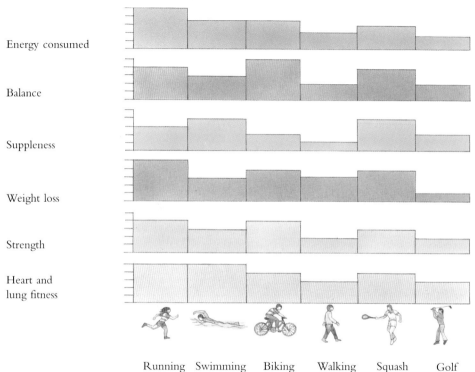

Energy consumed

Balance

Suppleness

Weight loss

Strength

Heart and
lung fitness

Running   Swimming   Biking   Walking   Squash   Golf

WHAT GOOD DOES IT DO?

This chart shows how different forms of exercise can bring different benefits. You can see, for instance, how running improves heart and lung fitness, but doesn't do as much for suppleness as, *for example, swimming. Biking, on the other hand, is great for developing your balance and co-ordination, and as good for building up muscles as running.*

Twitches — Slow-twitch red muscle (aerobic)

Oxygen & glucose in   Water & CO_2 out

Glycogen only — Fast-twitch red muscle (anaerobic)

Glycogen only   Excess lactic acid — Fast-twitch white muscle (anaerobic)

RED AND WHITE MUSCLE

Not all muscles get tired at the same rate. Nor are all muscles so good at providing instant power. This is because there are three kinds of muscle fiber: two kinds of red muscle and white muscle. A twitch is a single contraction of a muscle. One kind of red muscle twitches rapidly; the other twitches slowly. "Slow-twitch" red muscles cannot give a burst of rapid action, but can go on twitching slowly for hours without tiring. This is why muscles that have to keep going for a long time—such as your back muscles—are mostly slow-twitch. White muscles and fast twitch reds twitch powerfully but anaerobically (see p93) so get tired quickly.

Long term effects of exercise

What is regular exercise?

- Regular exercise is strenuous exercise lasting at least 20 minutes, three or more times a week.

Your body benefits from regular exercise in the following ways:

- Your resting heart rate goes down. The heart chambers get bigger, so your heart can deliver more blood at the same heart rate.
- Your ability to pump extra blood quickly when needed improves.
- Your heart rate goes back to normal more quickly after exercise.
- The muscles you exercise grow bigger. First muscle fibers grow thicker, then they increase in number.
- More enzymes are made in your muscles.
- Your strength, stamina, and endurance improve.
- Your ligaments and tendons become stronger. This reduces the chances of injury during any vigorous activity.
- Your joints become more flexible. This reduces the chances of spraining joints.
- You lose excess weight. Exercise does reduce weight—but it takes an awful lot of exercise to really make any difference if you're seriously overweight. Exercise must be combined with a controlled diet.

GLOSSARY

ABDUCTION A movement of a body part away from the midline of the body.

ABSCESS Pus cells made by the body's reaction to an infection.

ACUTE Medical conditions that start suddenly and last a short time.

ADDUCTION A movement of a body part towards the midline.

ALLERGEN Any substance that causes an allergic reaction.

ALVEOLI Tiny air sacs in the lungs.

ANEMIA A disorder in which blood is short of hemoglobin.

ANTIBODIES Proteins in the blood and other body fluids that fight infection.

AORTA The largest artery of the body, going from the heart's left ventricle and supplying oxygenated blood to all other arteries except the pulmonary artery.

APPENDIX A socklike structure at the start of the large intestine.

ARTERIOLE A small artery.

ARTERY A wide vessel carrying blood away from the heart.

ARTHRITIS Inflammation of a joint.

ATRIA The two upper chambers of the heart.

AXON Fine filament of a nerve that carries signals away from a nerve cell body.

AUTONOMIC NERVOUS SYSTEM The part of the nervous system that controls body processes such as heartbeat and breathing automatically, without you being aware of them.

BILE A greenish-brown fluid made by the liver and stored in the gallbladder that helps the digestion of fats.

BOLUS A lump of swallowed food.

BLASTOCYST The first group of cells forming from a fertilized egg.

BRADYCARDIA A slow heart rate.

BRONCHIOLE A small airway in the lung.

CAPILLARIES Narrow blood vessels that form a network throughout the body.

CARCINOMA A cancer of a surface layer such as the skin and large intestine.

CARTILAGE A rubbery substance that cushions and protects joints.

CEREBELLUM The area of the brain behind the brain stem concerned with balance and the control of fine movement.

CEREBRUM The largest part of the brain, made up of two hemispheres. It contains the nerve centers for thought, personality, the senses, and voluntary movement.

CEREBROSPINAL FLUID A nourishing and cushioning fluid that surrounds the brain and spinal chord.

CHROMOSOME One of the 46 sets of genes found in a cell nucleus —coils of DNA coated in protein.

CHYME Semi-digested food found in the stomach.

CILIA A tiny hairlike structure that waves back and forwards to move things along.

CNS Central nervous system (brain and spinal chord).

COLLAGEN A protein found in most body tissues that gives them elasticity.

COLON The upper part of the large intestine, extending from the cecum to the rectum.

CONGENITAL Present at birth. Congenital disorders may be inherited or may have occurred during pregnancy.

CORNEA The transparent layer that protects the front of the eye.

CORONARY Literally "crown:" arteries supplying the heart with blood.

CORPUS CALLOSUM The bunch of nerve fibers linking the two hemispheres of the cerebrum.

CORTEX The outer part, like a rind, of some organs, including the brain and kidneys.

CRANIAL NERVES Twelve pairs of nerves that are connected directly to the brain.

DENDRITE Tiny threads branching from a neuron body that receive impulses from neighboring neurons.

DENTINE The tough layer of tooth beneath the enamel.

DERMIS The inner layer of skin.

DIAPHRAGM The large domed sheet of muscle underneath the lungs.

DIASTOLE The resting phase of the heartbeat.

DNA Deoxyribonucleic acid. The giant coiled molecule found in each

cell nucleus that carries the genetic code.

ECG Electrocardiogram, a device to measure electrical activity in the heart.

EEG Electroencephalogram, a device to measure electrical activity in the brain.

EMBRYO A new human life for its first 8 weeks inside the womb when tiny organs are being formed.

ENAMEL The very hard coating of teeth.

ENDOCRINE GLAND A gland that secretes hormones directly into the bloodstream.

ENZYME A protein that speeds up a chemical reaction.

EPIDERMIS The outer layer of skin.

EPIGLOTTIS Cartilage flap at the top of the trachea that stops food from going down the wrong way.

ESOPHAGUS The tube that leads from the mouth to the stomach.

ESTROGEN A female sex hormone that stimulates the development of secondary sexual characteristics and prepares the lining of the uterus for a fertilized egg.

EXOCRINE GLAND A gland that secretes its products through a duct into a body cavity or onto the body surface.

EXTENSION The straightening out of a joint.

FALLOPIAN TUBE One of two open-ended tubes down which an egg travels from an ovary to the uterus.

FECES The semisolid waste products of digestion.

FETUS What an embryo becomes after 8 weeks of pregnancy.

FIBRIN A protein involved in blood clotting and wound healing.

FLEXION The bending of a joint.

GENE A unit of the genetic code giving the instructions to create a specific protein or structure.

GLAND A structure in the body that releases chemicals such as hormones and enzymes.

GLIAL CELLS Cells that nourish and support neurons.

GRAY MATTER The part of the brain and spinal cord that contains neuron cell bodies.

GUT The intestines.

HEMOGLOBIN Red blood pigment that carries oxygen to the tissues in red blood cells.

HEPATOCYTES Liver cells.

HIPPOCAMPUS A tiny structure in the brain that handles learning and long term memory.

HORMONES Chemicals sent out into the bloodstream to set off specific processes.

HYPOTHALAMUS A small structure at the base of the brain that handles the interaction between nerves and hormones.

ILEUM The last part of the small intestine that completes the absorption of nutrients.

INFLAMMATION Redness, swelling, and pain due to injury or infection as the chemical histamine stirs the body's immune system into action.

KILLER CELL A white blood cell that destroys damaged, infected, or malignant body cells.

KERATIN Tough protective protein found in hair, skin, and nails.

LACTEAL A small lymph vessel found in intestinal villi.

LIMBIC SYSTEM A ringlike structure in the brain that controls automatic body functions, emotions, and sense of smell.

LIVER A large organ in the abdomen that handles many vital chemical processes including preparing glucose for the body.

LIGAMENTS Strong, fibrous tissues that bind and support joints.

LYMPH A fluid that drains from body tissues into lymph vessels.

LYMPH NODES Lumps in the lymph system that filter germs from lymph fluid so that they can be attacked by lymphyocytes.

LYMPHOCYTES Small white blood cells that are involved in antibody production.

MARROW The soft inner part of long bones, where blood cells are made.

MEDULLA The soft, internal portion of some glands, organs, hair, and bones.

MEIOSIS The division of cells to create sperm and egg cells that involves halving the number of chromosomes to 23.

MEMBRANE A thin lining or covering layer.

MENINGES Three membranes that surround the brain and the spinal chord.

MITO-CHONDRION A small structure found inside cells where energy-producing chemical reactions take place.

MITOSIS The normal process of cell division that creates two identical copies of the original, complete with 46 chromosomes.

MOTOR NERVES Nerves that carry signals from the central nervous system to the muscles.

MYOFIBER Bundles of cells found in muscle fiber.

MYOFIBRIL Muscle-building blocks made up of two proteins, myosin, and actin.

NEPHRONS The tiny filtration units in the kidneys.

NEURON A nerve cell.

NUCLEUS The central control region of the cell.

OLFACTORY NERVE One of two nerves that take smell signals from the nose into the brain.

OPTIC NERVE One of the two bundles of nerves that carry visual signals from the eyes to the brain.

ORGANELLE One of the many tiny structures found inside body cells.

OSSICLE One of three tiny bones of the middle ear that transmit vibrations from the eardrum to the inner ear.

OSTEON A small unit used to build up compact bone.

OVARIES The pair of structures in the female abdomen that store eggs and release female sex hormones.

OVUM An egg cell. This turns into an embryo if it is fertilized.

PANCREAS A gland found behind the stomach that secretes digestive enzymes and hormones that control the level of glucose in the blood.·

PARASYMPATHETIC One of the two parts of the autonomic nervous system that controls day to day functions such as digestion automatically.

PARIETAL Referring to the wall of a body cavity.

PAROTID GLANDS The largest pair of salivary glands, on either side of the jaw.

PELVIS The ring of bone that makes up the hips.

PERICARDIUM The tough, fibrous bag around the heart.

PERISTALSIS The muscular contractions and relaxations that force food through the digestive tract.

PHAGOCYTES White blood cells that engulf and destroy invading organisms such as bacteria or a virus.

PINNA Ear flap.

PITUITARY GLAND The master hormone control gland, at the base of the brain.

PLASMA The clear, yellowish fluid and dissolved substances in which blood cells float.

PLATELETS The cell fragments involved in blood clotting.

PLEURAL MEMBRANE The sheet of tissue lining the inner surface of the chest cavity and the lung's outer surface.

PLEXUS A network of interwoven nerves or blood vessels.

PROGESTERONE A female sex hormone secreted by the ovaries and the placenta that prepares the lining of the uterus to receive a fertilized egg.

PROSTAGLANDINS Fatty acids that occur naturally in the body and act in a similar way to hormones, particularly involved in controlling pain.

PROSTHESIS An artificial replacement for a part of the body.

PULMONARY Referring to the lungs.

PUPIL An opening in the center of the iris through which light passes to reach the retina.

PULSE The echo of the heartbeat through the blood—the rhythmic expansion and contraction of an artery as blood is forced through it.

RECEPTORS Structures that sense things or pick up signals such as nerve impulses.

RED BLOOD CELLS Cells circulating in the bloodstream that contain hemoglobin and carry oxygen.

RESPIRATION Breathing and also the conversion of glucose and oxygen into energy in every body cell.

RETINA The light-sensitive lining at the rear of the eye.

SALIVA A digestive fluid secreted by the salivary glands in the mouth.

SCIATICA Pain caused by pressure on the sciatic nerve—usually felt in the buttock and thigh.

SCLERA The tough, outer white of the eye.

SENSORY NERVES Nerves that carry information from sense receptors back to the central nervous system.

SEPTUM A dividing wall such as the nasal septum.

SINUSOIDS Blood-filled spaces found in some tissues, such as in the liver.

SUPERIOR VENA CAVA The largest vein in the body, leading directly to the heart.

SUTURES Special joints that lock skull bones together so that they cannot move.

SYMPATHETIC NERVOUS SYSTEM One of the two parts of the autonomic nervous system. It prepares the body for action by, for instance, raising heart rate.

SYNAPSE A small gap at the junction between two neurons.

SYNOVIAL JOINT A joint lubricated by synovial fluid.

SYSTOLE The contraction stage of the heartbeat.

TENDON Strong, connective tissue that attaches muscle to bone.

TESTOS-TERONE The male sex hormone made in the testicles.

THYMUS A gland that makes and stores T-lymphocytes in childhood.

TRACHEA The windpipe.

TISSUE A group of cells or fibers that perform a similar function.

TRABECULAE Small bony struts that make up spongy (cancellous) bone.

TYMPANUM The eardrum.

UREA The waste from the breakdown of proteins and the nitrogen-containing component of urine.

URETHRA The tube that carries urine from the bladder to the outside.

URINE The clear, yellowish fluid waste from the kidneys.

UTERUS A hollow muscular tube in the female's lower abdomen. It becomes the womb when she is pregnant.

VAGINA The muscular tube from the uterus to the outside in females. This swells to become the birth canal in childbirth.

VAGUS NERVES The tenth pair of cranial nerves, controlling automatic functions such as heartbeat and digestion.

VALVES Flaps of tissue that prevent the backflow of blood in large veins and in the heart.

VEIN A blood vessel that carries oxygen-depleted blood back to the heart.

VENA CAVA One of the two large veins in the body that empty into the heart's right atrium.

VENTRICLES The two lower chambers of the heart.

VENULE A small vein.

VILLI Small projections in the intestinal wall that absorb nutrients.

VIRUS The smallest of all germs, a hundredth of the size of bacteria. They infect by taking over body cells.

WHITE BLOOD CELLS The colorless blood cells, including lymphocytes and leucocytes that protect the body from invading organisms.

WHITE MATTER The part of the brain and spinal cord that contains neuron axons.

X CHROMOSOME One of the two sex chromosomes. Women's body cells have two X chromosomes.

Y CHROMOSOME One of the two sex chromosomes. This makes a baby male.

ZONA PELLUCIDIA The tough, outer membrane or shell of a human egg.

ZYGOTE A newly fertilized egg.

INDEX

255

ACKNOWLEDGMENTS

The publishers wish to thank the following artists who have contributed to this book.

Kuo Kang Chen, Andrew Clark, Jeremy Gower, Sally Launder, Mike Saunders, Guy Smith, Steve Weston.

All photographs from the Miles Kelly Archive.